MES Guide for Executives:

Why and How to **Select, Implement, and Maintain** a Manufacturing Execution System

MES Guide for Executives:

Why and How to

Select, Implement, and Maintain

a Manufacturing Execution System

Bianca Scholten

Author	Bianca Scholten
Images	Bettina Logge-Mensing
Layout & cover design	International Society of Automation (ISA)
Translation	Grayson Morris

67 Alexander Drive
P.O. Box 12277
Research Triangle Park, NC 27709

Printed in the United States of America.
10 9 8

ISBN: 978-1-936007-03-5

Notice
The information presented in this publication is for the general education of the reader. Because neither the author nor the publisher has any control over the use of the information by the reader, both the author and the publisher disclaim any and all liability of any kind arising out of such use. The reader is expected to exercise sound professional judgment in using any of the information presented in a particular application.

Additionally, neither the author nor the publisher have investigated or considered the effect of any patents on the ability of the reader to use any of the information in a particular application. The reader is responsible for reviewing any possible patents that may affect any particular use of the information presented.

Any references to commercial products in the work are cited as examples only. Neither the author nor the publisher endorses any referenced commercial product. Any trademarks or tradenames referenced belong to the respective owner of the mark or name. Neither the author nor the publisher makes any representation regarding the availability of any referenced commercial product at any time. The manufacturer's instructions on use of any commercial product must be followed at all times, even if in conflict with the information in this publication.

Library of Congress Cataloging-in-Publication Data

Scholten, Bianca.
MES guide for executives : why and how to select, implement, and maintain a manufacturing execution system / Bianca Scholten ; [translation, Grayson Morris].
p. cm.
Includes bibliographical references and index.
ISBN 978-1-936007-03-5 (pbk.)
1. Computer integrated manufacturing systems. 2. Manufacturing processes--Data processing. 3. Production management--Data processing. 4. Factory management--Data processing. I. Title.
TS155.63.S3613 2009
670.285--dc22

2009020314

Foreword

When I first heard from Bianca about her plan to write a book about Manufacturing Execution Systems (MES), I was both excited and wary. She went on to articulate that it was her desire to illuminate some of the ambiguities surrounding this subject and clear up any lingering perceptions that MES is a black art. While I applauded her objectives, I was quick to remind her that others have tried this, and largely failed—not because they didn't understand the subject matter or how it could and should be applied, but because they apparently assumed that their primary audience is made up entirely of rocket scientists and that they all suffer from insomnia. My guess is that for the most part they were wrong on both counts (speaking for myself, I meet neither of these criteria). My challenge to Bianca was simple, though there was nothing simple about the goal she set for herself: write this book so that the average practitioner, whether from Operations or from IT, can understand it and benefit from it.

As the Chairman of the Board for MESA International (Manufacturing Enterprise Solutions Association), I have the privilege of interacting closely with many manufacturers that are planning to implement an MES solution, and I've noted a striking consistency through these conversations. It's clear that MES is that rare breed of solution that spans both the IT and Operations domains, and this brings about a whole set of challenges entirely distinct from the MES-related problem they're trying to solve; challenges associated with cultural and convergence-related conflicts rooted firmly in a lack of understanding of each other's domains. It's not hard to deduce that these "soft" challenges, if not understood and mitigated, will inevitably have a significant negative impact on the success of any MES initiative.

This is not a new observation, and these are certainly not new challenges. The six million dollar question is this: Will Operations and IT leadership learn that until there's clear alignment between their respective strategies, goals, and metrics, they and their company have little hope of successfully rolling out any systematic, high return manufacturing solutions? Another important question: How can everyone affected better understand the nuances of how MES implementations impact not just specific workflows and business processes, but how they drive the need for functional and cultural convergence?

I contend that the basic alignment of Operations and IT strategies, goals, and metrics, and the very necessary convergence between Operations and IT, will come about only through the education of both parties on 1) the functional components and activity models of MES, 2) their boundaries and touch-points with ERP, 3) the recognition that IT and Operations both bring essential capabilities and world views to this equation, and 4) that one without the other will not succeed.

This brings me back to Bianca's commendable goal of documenting this multifaceted sum of ideas, capabilities, and manufacturing experience in an easy-to-understand way. Her down-to-earth, anecdotal writing style is easy to read, and she has a refreshing knack for simplifying the most complex structures into clear, well organized content. I believe this book will go a long way towards educating decision makers at all levels and driving the convergence of Operations and IT that is so essential for manufacturers to succeed in this new economy.

Happy reading!

John Dyck
Chairman of the Board for MESA International

Contents

About the Author

Bianca Scholten is a principal at Accenture, a global management consulting, technology services and outsourcing company, with more than 223,000 people serving clients in more than 120 countries. She advises multi-nationals on the definition and realization of their manufacturing IT strategies. She is a voting member of the ISA95 committee. Ms. Scholten is also the author of the book *The Road to Integration; A Guide to Applying the ISA-95 Standard in Manufacturing*, for which she received the Thomas G.Fisher award (best new standards-based ISA book of 2007) and the Raymond D. Molloy award (best-selling ISA book of 2007). Two years later she won the 2009 versions of these same awards for *MES Guide for Executives*. Ms. Scholten received the ISA Standards and Practices Award for her outstanding contribution to the technical report *Using ISA-88 and ISA-95 Together*. She has published many papers in trade journals on subjects related to vertical integration and technical automation and has trained hundreds of professionals in applying the ISA-88 and ISA-95 standards.

Introduction

"Are you already working on a second book?" Dennis Brandl, editor of the ISA-95 Enterprise-Control System Integration series of standards, asked me during the ISA EXPO 2007 in Houston.

"Well, I'm thinking of writing a book about MES and ISA-95, for the boards of directors and management at manufacturing companies," I answered. "I've noticed that many of my clients have trouble demonstrating to management what a manufacturing execution system is and what it can do for you. The books and white papers currently available on that subject are all targeted to a technical audience. There's nothing for the management level."

"That's a good idea," Brandl said; then he noted, "It should be a small book, one they could finish on the plane."

I thought that was a good starting point. A book short enough that CEOs, CFOs, general managers, controllers, plant managers, production managers, and IT managers at manufacturing enterprises could read it cover to cover during a single flight, so to speak. Something that would give them a quick first impression of manufacturing execution systems; a book that answered questions such as "What is MES?," "How can MES help us improve our business results?," and "What can we expect when we start using MES?"

The following year, I further elaborated this plan. During my daily work as a management consultant, I collected examples, spoke with colleagues, and interviewed IT managers and production managers who've garnered a great deal of MES experience within their companies. This resulted in a wealth of information, the most important parts of which are summarized in this book.

"Why would we need a manufacturing execution system? We follow a SAP-unless policy," you may be thinking. Chapter 1 addresses the question whether an ERP system can provide sufficient support for plant processes, and if so, under what circumstances. Your company may not need an MES. In that case, you can put the book down after chapter 1.

But you might come to the conclusion after reading that chapter that your production processes are sufficiently complex that your ERP system doesn't adequately meet your plant's information needs. In that case, read chapter 2 to find out exactly what an MES is, and chapter 3 for a rough overview of the costs and benefits of such a system. If you see opportunities for your company, then chapter 4 will tell you in global terms how an MES selection project should proceed, and chapter 5 spells out what your company can expect during implementation and initial use of the MES. If your company has multiple production sites, it may be possible to use one and the same MES in all your plants; chapter 6 contains advice on developing and maintaining a multi-site MES template. Finally, chapter 7 gives a bird's-eye view of the ISA-95 series and how it can help both people and information systems communicate better, internally and externally, during your MES projects.

This book would not have been possible without the help and support of several organizations, companies, and individuals. My thanks go to WBF (The Forum for Automation and Manufacturing Professionals, formerly known as the World Batch Forum), whose yearly conferences offer professionals the opportunity to exchange information on best practices. Many WBF conference presentations have been a source of inspiration for this book. I also thank MESA (Manufacturing Enterprise Solutions Association International) for its congresses, white papers, and other sources of information that help manufacturing enterprises to reach a higher plane. I've quoted from several MESA studies and publications in this book. And thank you, ISA, for making it possible to develop useful standards such as ISA-95, and for publishing specialized books in the field of manufacturing IT.

I also extend my heartfelt thanks to all my interviewees; their real-world experience will help readers to understand what they can do to make an MES project successful. Thank you also, Gert-Jan van Dijk, Jos Hensen, Wouter Huijs, Jan Kelderman, and Sjoerd van Staveren, for reviewing the draft versions. Your critical comments and useful tips—based on your own practical experience—have made an extremely valuable contribution to the quality of this book.

And finally, of course, a big hug for my husband, family, in-laws, and friends; thank you all for providing an environment in which I found the relaxation, rest, and energy I needed to write this book.

Bianca Scholten

Rosmalen, the Netherlands, March 2009

CHAPTER **1**

We Don't Have an MES...

Someone thrust this book into your hand with the comment, "Read this! It's about MES." So now, of course, you're thinking, "Why should I? We follow a SAP-unless policy, so we don't need a separate MES package." But how exactly does that "unless" work? When is an ERP system sufficient for supporting activities on the shop floor, and what type of company is better served by a dedicated solution to provide information to factory personnel?

1.1 Modern Countries, Primitive Factories

In recent decades, industrial companies have invested much time and money in machine and production line automation on one hand, and in ERP[1] systems on the other hand. Between these two automation layers lies another, usually called the **MES layer**.[2] MES concerns the activities that take place within a manufacturing department. These include preparatory activities, such as detailed production scheduling and recipe management, but also retrospective activities, such as data collection, reporting, and analysis.

In many factories, the situation can be called primitive in regard to these activities. They use MS Excel for their detailed scheduling and reports, and MS Word to manage operator instructions and recipes. When there are advanced applications available, these come from various vendors and are not integrated. Figure 1.1 shows a typical example of the kind of stand-alone applications that factories use.

1 See the glossary for an explanation of acronyms used in this book.
2 For a more detailed explanation of the MES concept, see chapter 2.

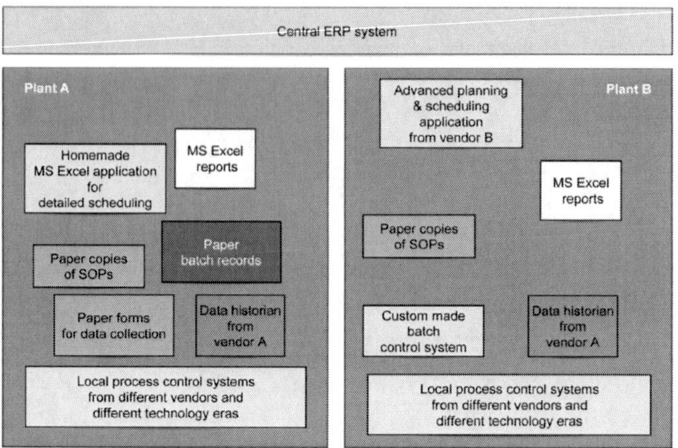

Figure 1.1 Typical example of the current level of automation in many factories

This outdated situation leads to many problems. For example, on the management and supervisory level, one has no insight into the current production situation. If the production manager wants to determine the source of a problem, he or she must first walk over to local systems in order to pull up the data files, then import these into a spreadsheet, and then reconcile the data with each other; only then—perhaps—will the answer to the question be revealed. This process can sometimes consume more than two entire workdays. It's impossible for the operator, supervisor, or plant manager to work proactively.

Management's lack of insight into the actual production situation also becomes painfully clear when you zoom in on scheduling issues. True, the ERP system does create a production schedule, but it isn't yet tailored to the actual capacity of the production lines, nor has it taken efficiency into account. That's why in nearly every factory, you'll find a scheduler who plans in Excel. These are people who carry around a great deal of knowledge in their heads. They really mustn't ever get sick or go on vacation, and they feel tremendously valued. The production department is completely dependent on this person. In the meantime, the plant manager's asking herself who's *really* the boss.

Once the schedule's complete, the supervisor prints out the production orders on paper, and places them in the proper order at the work stations. It remains to be seen whether this intended order will actually be followed. (Believe it or not, I've seen one of these stacks fly into the air when a door opened.) Employees on the production line have no insight into the internal dependencies among orders, and innocently threaten promised delivery dates. Another disadvantage of the current situation is that the operators can't concentrate on their most important task, namely, controlling the process. Operators spend hours copying data over from one system to the other. This is time-consuming and error-prone, and it results in data becoming available to other departments and systems only much later.[3]

By using a variety of nonintegrated systems, the problem of master data also rears its head. For example, if the factory's going to start using materials from new suppliers in the production process, or if it introduces new recipes, you have to update the master data in all those stand-alone systems. This is a time-consuming and error-prone process. If the master data conflict with one another, this can in the worst case lead to ordering the wrong raw materials or manufacturing end products that don't meet the specifications.

Many factories are accustomed to reporting average raw material consumption and production results back to the office after each production run, instead of the actual pounds or gallons consumed and produced. As a result—and depending on the type of industry—one must count or measure inventories on a daily or monthly basis and correct the ERP system's administrative inventory. Which means these factories don't really have their processes under control. With a little bad luck, the inaccuracy of the administrative inventory can

3 But note that this situation also has an advantage, namely, that operators have more "feeling" for the process and for the correctness of the data as a result of their hands-on involvement. This is a point to consider in implementing an MES.

even cause production stops because the proper raw materials aren't available. I don't have to explain to you just how expensive an hour or two of downtime is.

Come on, guys. This is the twenty-first century. There's got to be another way!

1.2 ...But We Have SAP!

1.2.1 SAP Unless...

Among the companies for whom I consult, more than 80 percent use SAP. Moreover, for reasons of standardization and reuse of efforts, and to limit maintenance costs, many opt for a **SAP-unless policy**. So when the production manager knocks on IT's door and asks for a specific scheduling system, the IT manager says, "No need. SAP can do that." And if the production manager then asks for a recipe management application, his colleague answers, "SAP can do that, too." The production manager has to have a silken tongue if he then wants to convince IT that a plant dashboard really isn't the same thing as a data warehouse.

Unfortunately, the incomprehension within manufacturing companies' walls reaches even further. Consider controllers who don't see why plant managers want detailed numbers. *We see from the monthly numbers that there's too much waste. How much more does she want to know? Surely she already sees what she needs to do!* And sometimes you have engineers who are instinctively against using SAP on the work floor—*And then I guess you'll want SAP to control the PLCs, too!*—but they lack the authority, communicative skills, and power of persuasion to alter the SAP-unless policy. In the end, the individual with the biggest mouth, or the one highest in the pecking order, gets his way. Unfortunately, that's not necessarily the person with the expertise to make the best decision.

In addition to my work as a consultant, I write a monthly article for the Dutch trade magazine *Automatie*. In early 2005, I wanted to write a piece about the degree to which SAP is or isn't suitable for supporting processes within factory walls. Toward that end, I visited two companies in Belgium: Helvoet Pharma and Agfa-Gevaert. Helvoet Pharma had deliberately chosen a SAP implementation in the factory. In contrast, Agfa-Gevaert chose not to use blanket SAP, based on the results of an internal inquiry into the pros and cons of SAP on the shop floor.

1.2.2 From the Trenches: Helvoet Pharma Chose SAP for the Factory Floor

Helvoet Pharma delivers rubber closures and aluminum and plastic caps to the pharmaceutical industry worldwide. At company headquarters in Alken, Belgium, I spoke with Herman Braeken, IT manager at Helvoet Pharma, and Wim Huybrechts, sales manager at SAP Belgium.[4]

> **HB:** We analyzed our automation needs some four years ago,[5] when SAP operated exclusively in the ERP market. We chose SAP for the ERP layer then, and for the SCADA layer, we chose Wonderware, with additional functionality for the MES layer. This way, we only needed to create an interface between SAP and the MES layer.
>
> Our strategy is to choose integrated solutions with as few interfaces as possible. We want to be FDA-compliant, because we supply to the pharmaceutical industry. That means that every interface has to be validated, which incurs substantial costs. For the interface between SAP and the MES layer, we considered Enterprise Application Integration tools. Then,

4 Herman Braeken (IT Manager Helvoet Pharma) and Wim Huybrechts (Sales Manager SAP Belgium), interview with the author, 2005.

5 This interview took place in early 2005.

suddenly, we heard that SAP had a module for the MES layer, the MO module. That was a big surprise for us.

WH: In 2004, SAP Belgium decided to throw the MO module into the spotlight. The module was almost ten years old, but it was rarely used because its user-friendliness left much to be desired. Over four years ago, SAP invested heavily in the development of the MO module, to breathe new life into it. Now it's on the rise. The solution has been heavily refined over the years, and the new developments can compete with specialized MES products.

A significant advantage, in particular for the pharmaceutical industry, is that all the functionality resides in one system, so that you have considerably fewer interfaces to validate. Electronic batch records are no longer spread across different systems. Moreover, the system is 21 CFR 11-compliant[6] and it can communicate with devices via OPC.

HB: Of course, every company has to analyze its own automation needs. The batch record features in SAP mesh well with Helvoet Pharma's needs. If you're willing to put in the effort, you can make it as detailed as you like. We've also experienced that OPC is possible using SAP. We've created a link between SAP and our scales, and these exchange information in two directions. In a later phase, we're going to link SAP to SCADA. For us, SAP is the only place left that contains master data. If we need a recipe, it's always sent from SAP to wherever it's needed.

WH: SAP isn't intended to control PLCs directly. It's common, however, to link SAP and SCADA systems together.

6 The FDA's 21 CFR Part 11 Rule on electronic records and signatures.

HB: In terms of stability, we've chosen to manage the SAP MO module centrally, in a redundant solution. We make three hundred batches a day with the mixer, so system availability is important. SAP runs at our site in Alken, and that's the central point for our other locations.

WH: You can also choose a decentralized solution. In that case, you put the SAP MO module on a separate server, possibly locally. In this situation, if the administrative system drops out, the production department doesn't come to a standstill.

HB: Our processes consist largely of manual labor. We've developed procedures so that the operators can always keep on working.

SAP is considered an expensive system, but that's relative. We already had SAP, so we didn't have to invest extra into hardware or software. We didn't have to buy a new package, and we didn't have to train people. The same people who first concentrated on the Sales and Distribution module now work on the other modules. Moreover, MES consultants aren't really cheaper than SAP consultants. There may be a slight difference compared with rates for SCADA or MES consultants, but not an appreciable one. And don't forget that consultant availability plays a role. A SAP consultant is easy to find, but just try to find someone who knows his way around WinCC![7]

1.2.3. From the Trenches: Agfa-Gevaert Chose SAP Where Possible, a Separate MES System Where Necessary

Agfa-Gevaert uses SAP worldwide. The Global Information and Communication Services department determines Agfa's IT strategy,

7 WinCC is Siemens' SCADA package.

and strives to put the production processes for all sites on one platform. This way, Agfa-Gevaert can put an end to its enormous number of separate systems and accompanying interfaces, whose maintenance costs a great deal of time and money. Agfa's starting point is thus: SAP, if possible.

During my trip to Belgium, I spoke with Marc Verhaegen, Manager of International Projects and IT Architect within Agfa-Gevaert's GICS / Manufacturing Services group.[8]

> About two[9] years ago, we researched whether the strategy we had in mind was suitable for all our sites. To this end, we compared several of our locations in Europe. The degree of complexity in production processes was the determining factor in our decision whether to implement SAP alone, or to provide an extra MES system.
>
> In Mortsel [Belgium], the processes turned out to be so complex that we couldn't manage them with SAP alone. In Mortsel, film for medical and graphical applications is manufactured in master rolls (up to sixty-seven inches wide and five miles long) in a continuous process with speeds of more than 330 yards per second. These rolls have to go through multiple production steps, coatings must be applied to one or both sides of the material, in the dark, and one roll can result in several rolls or vice versa. It's exceptionally important to be able to follow the genealogy of these rolls through the different production steps. A roll error that arises in a particular step of the production process is guaranteed to show up in a different place after the roll has undergone its last processing step. Error tracking is essential in order to be able

8 Marc Verhaegen (Manager of International Projects and IT Architect, Agfa-Gevaert, GICS / Manufacturing Services), interview with the author, 2005.
9 This interview took place in early 2005.

to optimally cut the roll into end products. That way, we avoid wasting expensive raw materials, such as silver.

Moreover, it requires particular attention to get the rolls to the right work center at the right moment. We've solved this by constantly tracking the situation on the work floor, and feeding this back to the transport department via an integrated MES system. We've linked our process control systems using robust and user-friendly interfaces that support the operators' manual interventions. The detailed scheduling in our textiles department also requires a custom solution. Supporting all these complex issues with standard SAP R/3 is, in our view, impossible. And thus an MES system was the best solution for the Mortsel site.

The choice to use a separate MES system does have a few disadvantages. For example, you can't avoid double data storage. Certain information—like (semi)finished product inventory—arises in the MES layer. SAP needs these same data, but for completely different purposes, such as calculating the production price and creating financial reports. We've decided not to send all details from the MES to SAP, but to aggregate the data.

Interfaces are a constant concern, and data reconciliation is crucial. That's why we choose a single system wherever possible. At our sites in China and Leeds, we use SAP exclusively.

From our study, we drew the conclusion that—in our sector—links between SAP and process control systems, which are intended to drive processes, can be avoided. We've had some bad experiences with this within Agfa-Gevaert. In our opinion, environments with complex dynamic processes are better supported by an MES system, because of its simple user interface, the required flexibility for data collection and data manipulation, and the specific automation and support needs characteristic of our industry. Requirements concerning a shop floor control system's availability and independence can also argue in favor of a separate MES system.

At the moment, we aren't ruling out a link between SAP and control systems, but only where non-time-critical data are concerned (such as canceling a production order), and always under the control of the workers on the floor.

1.2.4 But What if SAP Can't Provide a Suitable Solution?

These interviews reveal that opinions are divided on the suitability of SAP for the MES layer. In any case, it's best not to blindly follow a SAP-unless policy. Be on your guard as soon as the subject of automating typical MES activities comes up. The following table is a tool you can use[10] in deciding on a SAP-unless policy.

Table 1.1 Considerations when choosing whether to follow a SAP-unless policy

Greater likelihood that traditional ERP functionality is suitable for the MES layer	Lesser likelihood that traditional ERP functionality is suitable for the MES layer
Simple processes, stable routings	Complex processes, unstable routings
Long production runs (days, weeks)	Short runs (minutes, hours); many batches (for example, more than five batches per day)
One-to-one correspondence among production orders from office to factory	Splitting or merging production orders in the factory
Manual activities, driving and registering operator activities	Time-critical integration with the process control layer

10 The arguments here are largely derived from the Agfa-Gevaert study.

> Beware!
> The world is in motion. SAP and other traditional ERP vendors buy companies, thereby gaining access to applications developed as dedicated MES solutions. These applications can be worthy competitors to the MES solutions from other vendors. The boundary between *vendors* blurs. That's okay. The boundary between ERP *applications* and MES *applications*, however, should never fade. Why will become clear in the following chapters.

You might have concluded that the processes in your factories are so complex that your central ERP system can't support them. You need a suitable solution. And MES vendors claim they can provide it. What is MES? And what functionality can you generally expect from an MES package?

CHAPTER **2**

MES...? What's That?

Some twenty years ago, pioneers began creating custom software for operators, supervisors, schedulers, plant managers, engineers, maintenance mechanics, and other factory personnel to provide functionality missing from ERP systems and process control systems. Slowly but surely, more and more software vendors discovered this hole in the market. They began developing standard solutions, to which they've kept adding more and more functionality over the years. In the early nineties, the term manufacturing execution system was coined for this type of solution. Which company activities do such IT applications support? Are they suitable for every kind of industry? And what can an MES do that your average ERP system can't?

2.1 MESA and ISA

The organizations MESA and ISA lead the way in providing valuable information about manufacturing execution systems, such as white papers, technical reports, standards, and models and terminology. Let's take a closer look at the explanations these organizations give for the phenomenon *MES*.

2.2 MES According to MESA

Jan Snoeij, chairman of MESA's European Board of Directors, explains what **MESA** is and does.[1]

1 Bianca Scholten, "Plant to Enterprise; MESA organiseert Europese conferentie in Utrecht," *Automatie* (no. 9, 2007), 10.

MESA International stands for *Manufacturing Enterprise Solutions Association International.* It's a nonprofit organization. MESA was founded in 1991 to exchange knowledge, experience, and best practices for using MES among vendors, system integrators, analysts, and manufacturing company personnel.

The term MES arose in 1991. Multiple people claim they invented it. At that time, it stood for *manufacturing execution systems.* Today, it means *manufacturing enterprise solutions.* After all, MES is more than just a system for production control. Issues such as quality, inventory, maintenance, product data management, and product life cycle management can't be viewed as separate from the MES domain. ... That's why we changed the term in 2004.

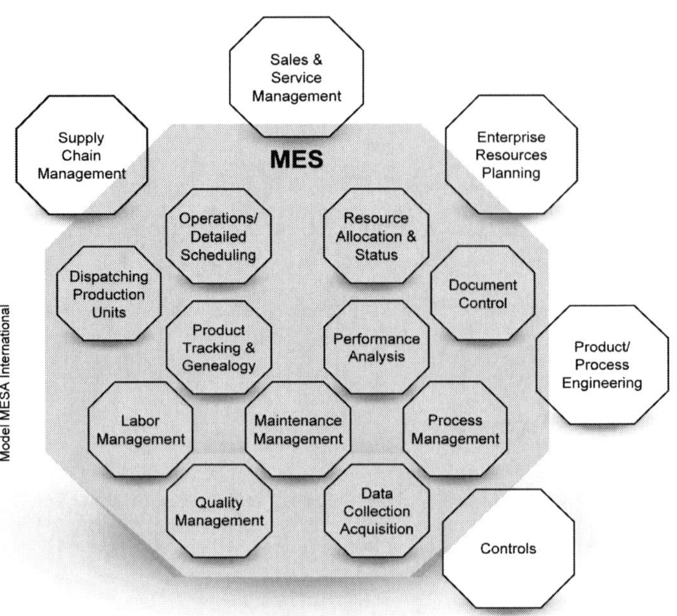

Figure 2.1 MES according to MESA White Paper # 02

In one of its first white papers,[2] MESA distinguished eleven manufacturing execution activities, which later gained recognition primarily thanks to the MESA honeycomb model (see figure 2.1). The eleven activities are Resource Allocation and Status, Operations/Detail Scheduling, Dispatching Production Units, Document Control, Data Collection/Acquisition, Labor Management, Quality Management, Process Management, Maintenance Management, Product Tracking and Genealogy, and Performance Analysis.

Simply put,[3] the original concept **manufacturing execution system** concerns information systems that support the things a production department must do in order to

- prepare and manage work instructions;
- monitor correct execution of the production process;
- gather and analyze information about the production process and the product, and feed this back to other departments, such as accounting and logistics;
- solve problems and optimize procedures.

MESA's honeycomb model gives a good first impression of the scope of MES.[4] It's a useful communication tool for discussing MES and related issues in broad terms. The advantage of the model is that it's very clear. But if you want to know what the relationships are among the various activities, and what information they exchange with one another and the outside world, then ISA-95 series is a more useful tool.

2 MESA International, *White Paper #02: MES Functionalities & MRP to MES Data Flow Possibilities* (Chandler, AZ: MESA, 1997).

3 Of course, this is a rather superficial explanation. For a more detailed description of MES, you can consult the MESA white papers and ISA-95 Part 3.

4 MESA also provides more extensive models that cover the broader concept of *manufacturing enterprise solutions* rather than the original *manufacturing execution systems*. (See appendix A.) However, this book focuses solely on a first acquaintance with MES in the traditional sense of *manufacturing execution systems*.

2.3 MES According to ISA

ISA stands for *International Society of Automation*. This nonprofit organization defines standardization in the field of industrial automation as one of its primary objectives, in addition to certification, education and training, publications, conferences, and shows. Examples of familiar and popular ISA standards are ISA-84 (*Functional Safety: Safety Instrumented Systems for the Process Industry Sector*), ISA-88 (*Batch Control*), and ISA-99 (*Manufacturing and Control Systems Security*).

ISA-95 bears the title *Enterprise-Control System Integration*. ISA-95 is not an automation system, but rather a method, a way of working, thinking, and communicating. The method is described in several documents, each about a hundred pages long. The documents contain models (figures) and terminology you can use to analyze an individual manufacturing company. All the models are focused on specific aspects of ERP-MES integration; they each throw light onto the issue from a different perspective.

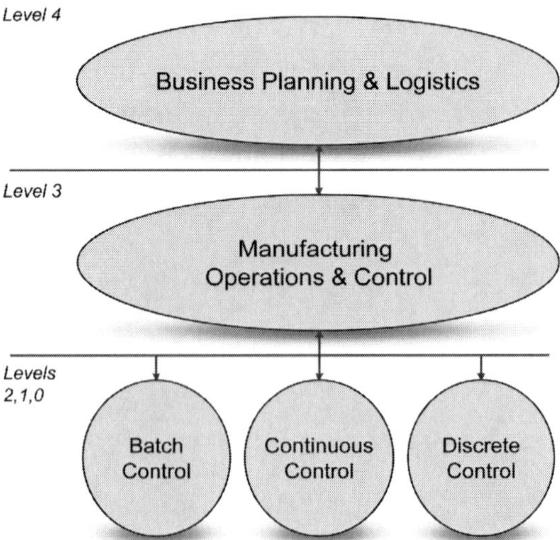

Figure 2.2 ISA-95 functional hierarchy model

By the way, the ISA-95 series[5] never mentions the terms *ERP* and *MES*. Nor should standards like this tell vendors what they may and may not sell. What ISA-95 does do is mark the boundaries between different levels of decision-making, in which different types of information belong. The Functional Hierarchy model (see figure 2.2), for example, defines a **Level 4** that's focused on the longer term (months, weeks). This is where decisions get made on ordering materials, sending invoices, long-term production and maintenance scheduling, and the development of new products. Level 4 is popularly called the **ERP layer**.

In contrast, **Level 3** focuses on a somewhat shorter term (days, hours, minutes). Decisions here have a direct influence on the plant's efficiency, product quality, material storage locations, and machine availability, among other things. People usually call this the **MES layer**. All the functions MESA defines in its honeycomb model belong on ISA-95 Level 3. Beneath this level are **Levels 2**, **1**, and **0** (with a time frame of minutes, seconds, or milliseconds), where the production process itself takes place with the help of sensors, PLCs, SCADA solutions, and other types of control systems.

During the selection and implementation of MES and related systems, manufacturing companies are confronted with an overlap in functionality among the packages the different vendors offer. "What's the best package for which requirements?" they ask themselves. And if they don't have any tools to help them decide, they're likely to make unfortunate choices. That's how some businesses that have carried their SAP-unless policy as far as they can, are now saddled with an unnecessarily expensive and inflexible IT landscape. They've interwoven their ERP and MES layers like spaghetti. As a result, when they buy a new plant, they have to untangle the entire ball of spaghetti—even if the new plant has its own, well-functioning MES system that could very easily have been integrated with the ERP

5 Often abbreviated S95.

layer if the boundary between them had been chosen more logically. ISA-95 helps companies avoid these kinds of cost-inefficient choices.[6]

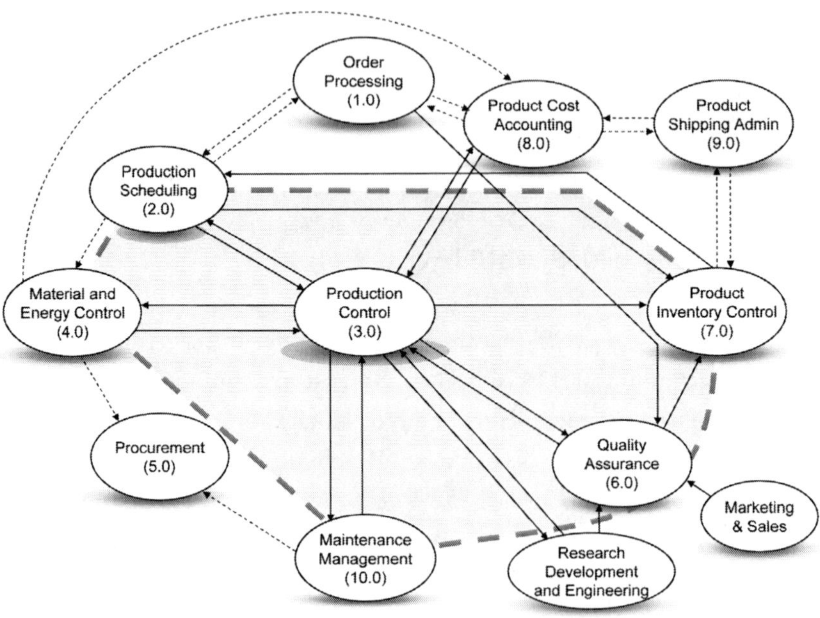

Figure 2.3 ISA-95 functional enterprise – control model (simplified)

The Functional Enterprise-Control model in ISA-95 (see figure 2.3) clearly shows where that boundary lies, logically speaking. It shows us twelve functions you'll encounter in every manufacturing company, though each company may give them different names and assign the functions to different departments. And ISA-95 certainly doesn't intend to prescribe an organizational structure with this model. The model only clarifies which functions are and are not involved in integrating ERP and MES systems. The fat dotted line represents the boundary between the Enterprise Domain and the Control Domain. Everything outside the boundary lies in the **Enterprise Domain**

6 See chapter 7 for a more detailed explanation of ISA-95.

(read: ERP), and everything inside it lies in the **Control Domain** (read: MES and the lower levels, such as SCADA).

Now, several functions straddle this border. Take a look at the function Production Scheduling. It's immediately clear that you'll have to pay special attention to determining the boundary between ERP and MES here. For there's a type of production scheduling that belongs on Level 4, but also one that belongs on Level 3. So be sure to distinguish between Level 4 scheduling and Level 3 scheduling. The first endeavors to "deliver the right product in the right amount and the right quality to the right customer at the right time" (*effectiveness*). The second endeavors to make the most *efficient* use of resources possible within the bandwidth of the master production schedule, for example by combining orders and thus limiting changeovers and cleanings to a minimum. The ISA-95 series offers even more detailed models that clarify more precisely where the exact boundary between the two lies.

The same applies to inventory information. The functions Material & Energy Control and Product Inventory Control (see figure 2.3) also straddle the boundary. Here, too, it's important to make good choices when assigning functionality to either the ERP system or the MES system. Level 4 contains the inventory strategies with associated order parameters such as series size and safety stock levels. Level 3, on the other hand, is primarily concerned with the physical handling of the materials. What's where? What's the expiration date for this raw material? Has the lab already released this lot? In short: this is an entirely different level of decision-making, home to entirely different information. But note that various traditional ERP vendors offer warehouse management functionality on Level 3, and these can certainly be suitable solutions. Here, too, make sure the two types of information aren't intertwined, by using separate, independent modules, for example. That will keep you flexible for the future.

According to ISA-95, Quality Assurance and Maintenance Management also contain decision-making on both Level 4 and Level 3.

For a more detailed description of the activities within Level 3, ISA-95 provides the following model (see figure 2.4).

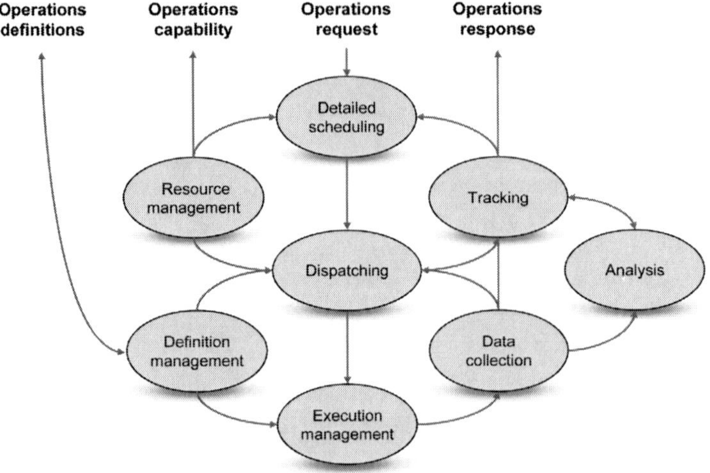

Figure 2.4 ISA-95 generic activity model of manufacturing operations management

This ISA-95 model clearly and logically organizes all the activities within the Control Domain. You can apply it to analyze a production department, a maintenance department, a laboratory, or a warehouse. Toward that end, someone within the production department must maintain information on the availability of production personnel, machines, and materials (Resource Management). It's also necessary to maintain the recipes, SOPs, and assembly instructions that employees use (Definition Management). A detailed schedule is required in order to combine orders optimally, taking into account the limited capacity of production lines, changeover times, and cleanings. Then someone has to send out the work orders and assign the tasks to shifts (Dispatching). The Execution Management function ensures that production personnel do indeed carry out those tasks, in conformance with the applicable quality standards. (Note: actual production execution takes place on a lower level.) Finally, during the production process someone must gather various data (Data Collection) and transform them into information (Tracking) to be used for tracing and genealogy, and for optimizing the production department's activities (Analysis).

The activities of maintenance departments, laboratories, and warehouses follow this same model.

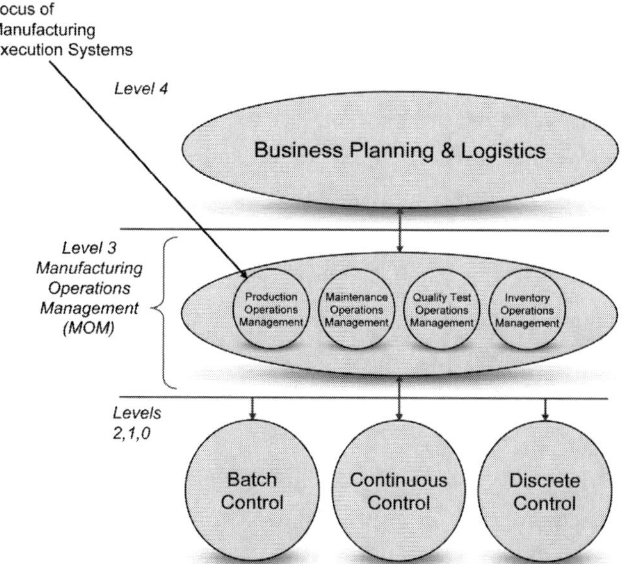

Figure 2.5 Manufacturing Execution Systems offer functionality that supports ISA-95 'Production Operations Management' activities, which are part of the Manufacturing Operations Management (MOM) domain

In summary, ISA-95 defines a Level 3, where not only the plant's activities take place, but also those of the maintenance department, the laboratory, and the warehouse. ISA-95 calls the activities of these four departments together Manufacturing Operations Management (see figure 2.5). This book focuses on the activities within the plant and does not further discuss maintenance departments, laboratories, and warehouses.

2.4 What Exactly Do MES Vendors Have to Offer?

Over the years, many departments within manufacturing enterprises have built their own Excel and Access databases, and they've become quite attached to them. The applications are fully attuned to the needs of their users, and they're easy to change and expand. However, the systems aren't integrated, and they're poorly documented. All the knowledge is carried by one person, or just a few people. This puts the plant's continuity at risk.

The ideal situation would be to replace all those stand-alone databases and applications with a single integrated MES system. In practice, however, nearly all MES vendors specialize in just a few of the MES activities. Some offer advanced reporting tools, but provide absolutely no scheduling functionality. Conversely, there are specialized Advanced Planning & Scheduling vendors that don't provide any reporting. Many vendors are busy expanding their functionality so that they cover all the MES functions, but aside from a few pioneers, they aren't there yet. At present, the chance is great that you won't succeed in getting all that functionality from one and the same vendor without paying for it in terms of quality. Fortunately, a "best of breed" approach is becoming easier and easier to realize, but in that case be prepared for cost-inefficient interfaces and problems managing master data.

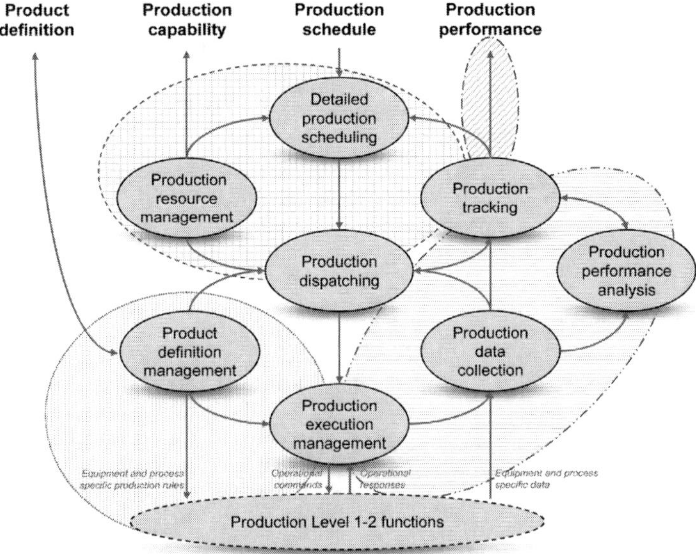

Figure 2.6 MES vendors offer modular software, the boundaries in these modules usually don't coincide with the activities that ISA-95 defines

Most MES vendors offer their functionality in modules. The cross-hatched areas in figure 2.6 represent MES modules that can be independently purchased and implemented. As figure 2.6 makes clear, the boundaries of MES modules don't usually coincide with those of the activities ISA-95 defines.[7] Broadly speaking,[8] vendors offer modules for the following MES functionality (see figure 2.7):

- Detailed production scheduling
- Product definition management and production execution management
- Historians (including an interface with Level 2 systems)
- Diverse reporting and tracking & tracing modules, covering (for example) quality, efficiency, and genealogy

7 The ISA-95 Production Operations Management model provides a vendor-independent definition of the activities within a plant, independently of whether these are supported by an information system.

8 This is a highly simplified view of the complex MES market, solely intended to give a first impression.

- Dashboards
- Workflow management
- Interfaces, integration with ERP
- Plant model

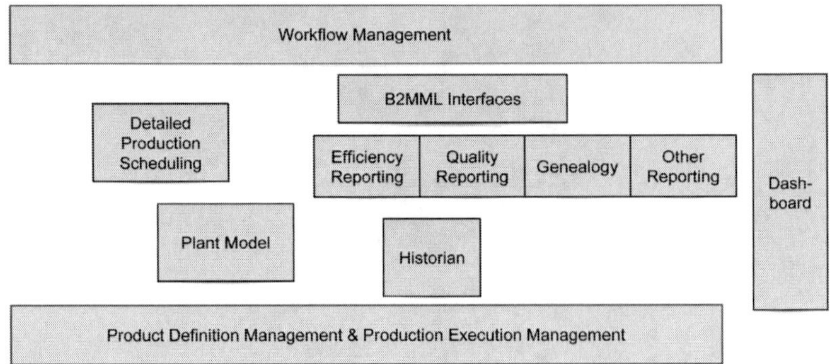

Figure 2.7 Simplified, vendor-independent overview of modules offered on the MES market

Below is a short, global description of the functionality these kinds of modules provide.

2.4.1 Detailed Production Scheduling

Detailed production scheduling software vendors offer specialized functionality for detailed and finite capacity scheduling. Some solutions are simple replacements for Excel applications. They help the scheduler to quickly compare alternative schedules. Others are more advanced. Here are a few examples of what a detailed scheduling solution can provide:

- Inventory scheduling: give short-term (weekly, daily) insight into the final product inventory to be produced and the required raw material stocks
- Schedule production employees, taking into account the required skills per line

- Efficiency scheduling: very quickly calculate optimal orderings and combinations of production orders and interim steps. In doing so, the software takes into account optimal cleaning and changeover times
- Simulation functionality: create and compare multiple schedules
- Capacity scheduling: provide insight into the available capacity in terms of days up to a few weeks, to facilitate synchronization with the sales department
- Automated interface with Level 4 systems
- Create a detailed schedule that can be sent automatically to the production department / work stations, if desired
- Send customer-specific data along with the production orders
- Support for the JIT (just-in-time) strategy; scheduling the timely acquisition and removal of materials

Do note that a scheduling package itself almost never makes decisions. It works like a navigation system: it gives the scheduler advice, but he or she is still the one behind the wheel. After all, there are plenty of reasons to deviate from the ideal schedule. Consider an important customer who should be given priority, or reducing commotion on the shop floor.

2.4.2 Product Definition Management and Production Execution Management

Many MES vendors offer modules to manage information about how the product should be manufactured. For example, modules may cover the development and management of recipes, assembly instructions, flow charts, and SOPs. It's important that the solution you choose closely meshes with your industry type. For example, batch packages are primarily suited for managing recipes for batch processes; they'll probably let you down if you also want to define packaging compositions, and they're less suitable for continuous processes.

Before you buy one of these modules, it's also important to clearly delineate the characteristics of your production processes. Do you require a high degree of flexibility? Is version control important? Must the module link in to the process control systems? And there are many more points to consider that you should be aware of before you go shopping.

Broadly speaking, these types of modules offer one or more of the following functionalities:

- Central management of master recipes / instructions
- Generation of order-specific instructions (such as control recipes)
- Defining, and if desired, managing and monitoring process steps
- Maintenance and feedback of order status
- Ability to quickly and easily develop or change recipes / specifications, and to bring these into production quickly and correctly
- Version control
- Material master data management
- Electronic signatures (compliance with 21 CFR Part 11, particularly relevant for the pharmaceutical industry)
- Automatic downloading of product-specific machine settings to SCADA or a PLC
- An automated interface with the ERP system for BOM synchronization

2.4.3 Historians

Historians are "flat" databases, in contrast to "relational" databases. Relational databases are built on logical structures, in order to preserve the internal relationships among the data. An important disadvantage of relational databases is that the structures make the system slow. That makes them less suitable for real-time gathering and long-term archiving of massive quantities of process data, such as temperatures, pressure readings, start times, stop times, speed measurements, alarms, and so on.

In contrast, historians use a variety of storage and retrieval strategies (such as compression) to realize the required performance. They've been specially developed to collect, retrieve, and store enormous quantities of data in a fast, simple, and cost-effective manner. MES vendors usually offer historians as independent modules, which then form the basis for extensions such as reporting modules. Historians have automated interfaces for collecting data from various sources such as PLCs, SCADA, batch systems, and so on. My colleagues liken them to a kind of vacuum cleaner.

2.4.4 Reporting and Tracking & Tracing Modules

MES vendors that offer historians usually also sell reporting modules. There are also dedicated vendors. Often, these modules are limited to standard reports such as OEE, quality reports, shift reports, daily reports, weekly reports, electronic batch records, efficiency reporting, and so on. Experience has shown that end users usually aren't satisfied with these standard reports. Unfortunately, it only becomes obvious after purchase that people would rather use their own company-specific OEE calculation, or generate a shift report with company-specific content.

The MES market isn't yet advanced enough to offer subsector specific reports on a large scale. So when you purchase and implement reporting functionality, assume you'll incur extra costs for configuring and / or custom–programming the reports. Manufacturing companies can keep their costs low by standardizing reports internally (to the extent possible) across the various departments/plants.[9]
The ability to compose your own reports without programmer intervention still leaves much to be desired in many MES solutions.

9 See chapter 6 for more information on how you can develop a multi-site MES template.

Tracking & tracing is also an important functionality in many sectors; for example, the government or the company's customers require it. Tracking & tracing traverses the entire supply chain, from the original raw material up to and including the final end product. MES plays a role here, because during the production process, semifinished products are often combined or split apart. Level 3 is responsible for collecting and storing the genealogical data within the scope of the plant. Many MES vendors offer functionality to do this.

2.4.5 Dashboards[10]

Supervisors, plant managers, and other plant employees also need tools to be able to analyze various situations ad hoc. Most MES solutions can, at the very least, export a variety of data to Excel. Some vendors offer more advanced and dedicated analysis tools, such as for root cause analysis, Pareto analysis, and so on. Plant dashboards are also analysis tools.

A **plant dashboard** provides real-time and historical information and analysis options to diverse employees, such as operators, plant managers, maintenance staff, quality staff, shift supervisors, technologists, and so on. A dashboard doesn't necessarily need its own database. A dashboard is a window within which the employee can view different systems or modules from a single screen (see figure 2.8). For example, employees may see an order's progress status, or a line's OEE. They also have access to other systems in order to request ad hoc information.

10 Dashboarding is a typical example for which SAP offers dedicated functionality. This makes this traditional ERP vendor a fully worthy competitor to traditional MES vendors.

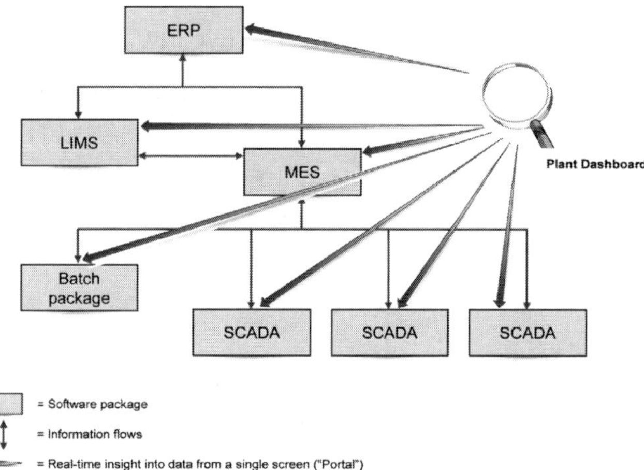

Figure 2.8 Sample implementation of a plant dashboard

I asked Tim Meijer, at the time a consultant with Ordina TA, what the difference is between a plant dashboard and a company-wide Business Intelligence (BI) system.[11]

> MESA presents[12] a kind of "Christmas tree model" that consists of layers of performance indicators and analysis indicators.
>
> Consider a manufacturing company with multiple sites. Profit is a performance indicator for the company as a whole. If profit deviates from expectations, you go looking for the cause. It might turn out that things aren't going well at one of the sites. So a performance indicator for a particular layer (in this case, the site) is an analysis indicator for the layer above it.
>
> Within a manufacturing company, production efficiency is one of the factors that influence profit. Other factors might include market demand, inventory size, pricing levels, and so

11 Bianca Scholten, "Plant Dashboards; Veel fabrieken sturen nog via de achteruitkijkspiegel," *Automatie* (no. 2, 2008), 2.

12 MESA International, *MESA Metrics that Matter Guidebook and Framework* (Chandler, AZ: MESA, 2006).

on. Business Intelligence starts at the top, and production is just one of the facets. Production is located more or less in the bottom layer of the 'Christmas tree' in BI systems.

Also, a characteristic difference between Business Intelligence systems and plant dashboards is that BI systems refresh their data at a much lower frequency—for example, once per day, or once per half-day, or once per shift. In the meantime, however, the operators determine what happens in the plant. With BI, they don't get any feedback until afterward. In a factory, you want to know how the plant's working, *now*. If you define your performance indicators well, then the operators will be looking at the same things during the process, and adjusting them right at the moment they need to do so. Business Intelligence systems aren't set up for that.

2.4.6 Workflow Management

The production department manufactures products. In doing so, operators make use of instructions (for example, recipes and SOPs). But what if exceptions occur? And how do you synchronize activities that involve multiple people and multiple systems and departments? Workflow management systems offer a solution for this. These systems weren't originally developed specifically for manufacturing companies, but MES vendors have discovered the concept, and they see the advantages for their customers. The functionality goes further than just sending an e-mail, for example, to the quality department, which has to evaluate a batch before it can move to the next step. The disadvantage of e-mail is that everyone can send it everywhere and no one manages follow-up. However, workflow management systems do take into account authorizations, event follow-up, and accept giving people room to choose their own solutions within a certain bandwidth, and in the meantime keeping the total process in view and under control.[13]

2.4.7 Interfaces with ERP

Level 4 (the ERP level) doesn't just send the master production schedule and BOM information to Level 3; it also wants to get information back. Many MES vendors offer standardized interfaces for all of this (B2MML[14]). Using one of these, the MES solution might, for example, send the following data back to the ERP system:

- Material efficiency and the actual raw materials consumed, including amounts (for accounting and also for accurate inventory control on Level 4)
- Labor efficiency, personnel usage, and so forth
- Production volumes, percentage rejected
- Data for calculating maintenance costs, energy and environmental costs, and waste stream costs
- Maintenance requests

2.4.8 Plant Model

The basis for every MES application is a model of the production process and its associated information. If you buy modules from different vendors, each module will have its own plant model. That has disadvantages.

Technically speaking, it's becoming easier and easier to develop interfaces between systems and thus exchange a variety of information. But if those systems all have their own views of the process, then synchronizing the information in the different systems is fundamentally complex, and in the long term a constant source

13 Bianca Scholten, "Productie als schakel in de keten," *Automatie* (no. 4, 2008), 20.

14 See chapter 7 for an explanation of B2MML.

of concern. With the arrival of the ISA-95 series of standards, the market now has access to standard data models that manufacturing companies can use to model their processes and information clearly and unambiguously. This means applications can jointly use a single, shared database. A few MES vendors have recently started offering solutions based on ISA-95 that solve this problem.

Section 2.4 has presented just a few examples of what the different modules can offer. It isn't a complete list. You can find more detailed information in the vendors' product sheets.

So the MES market offers quite a lot of functionality. Often, however, you don't need your information system to support all the Level 3 production functions. Certain functions have priority because you expect them to provide many advantages. For other functions, an advanced module may be overkill. Where priorities lie greatly depends on the manufacturing company's strategy, and the characteristics of the production processes. Your investments thus depend on the business case. But what does MES actually cost? And what advantages does it bring?

CHAPTER **3**

What Does MES Deliver? And What's It Going to Cost Us?

MES comes with a price tag. And since you can only spend your money once, you have to make choices. Should we buy an MES? If so, which modules are worth the investment? How will the MES functionality help our company to become more successful? Are there market numbers on what MES has done for our competitors? And what it's cost them?

3.1 Who Believes in MES?

In many automation projects, it's common practice to spell out the expected advantages. Sometimes people go so far as to calculate a Return on Investment (ROI). To do that, however, you have to be able to make a fairly accurate and realistic estimate of the costs and the benefits. In the case of MES projects, I almost never see anyone put extensive thought into this beforehand. Why not?

"They didn't ask us to do it for SAP, either," one of my clients told me. But that's not a good reason not to do it for an MES, of course.

Every time I lead the preparation of an MES project as a consultant, step one is to specify the expected advantages with the board and management. Many MES projects concern implementing functionality for reporting and analysis. Companies are looking for more insight into the way things are going in the plant, as a basis for proactive decision-making and continuous improvement. In that kind of project, it's impossible to determine beforehand how many dollars in savings or profit the new insights from the MES will add to the bottom line.

Even companies that already have an MES usually can't produce hard numbers on what it's done for them. Often they list concrete savings and improvements, but they can't fully ascribe these to the MES with total certainty.

At MESA's European conference in 2007, the organization asked the audience several questions. They'd handed out green, red, and yellow cards that stood for the answers *Agree*, *Disagree*, and *No opinion*, respectively. One of the questions was, "Who believes that MES can produce significant advantages for industrial companies?"

I was in the audience, and I thought to myself, "Hmmm. *Who believes*... Apparently, MES is a belief." By the way, most of the audience held up a green card. Believing isn't the problem. But wouldn't it be great if plant managers and IT managers could walk up to their bosses with hard-and-fast numbers?

3.1.1 From the Trenches: Arla Foods' MES Business Case

In September 2008, I interviewed Arne Svendsen, head of Manufacturing Services & Automation at Arla Foods Global IT, about Arla's MES project.[1]

> The initial business case for our MES project was one of the business units. They needed the ability to collect data on the physical process of making cheese. They didn't have any connected tools and they wanted to get the process data into an easily usable form.

1 Arne Svendsen (head of Manufacturing Services & Automation, Arla Foods Global IT), interview with the author, September 2008.

We did not make a comprehensive calculation for the business case. The production director for that business unit was visionary. Optimization of the manufacturing process was the driver for the first case; that was in 2004. I visited twenty-five of Arla's sites to "sell" the idea of collecting the data, which could then be used for many different purposes.

It's better not to create a comprehensive business case, but rather to focus on the strategic advantages. That will open eyes at the level of production management. What we're doing with our MES is bringing business agility to the manufacturing space. You need a strong Manufacturing-IT foundation so that whenever you want to adapt to a new situation, you have the basic infrastructure to do so. That's the promise of MES. The market changes all the time, and an MES makes it possible to introduce new products right away, including adjustments to printing, labeling, and so on.

We started with a focus on process optimization; then we added downtime optimization, and after that we started to collect all kinds of data for analyses. The data is now available online. This has resulted, for example, in a 10 percent improvement in line efficiency over the last twelve months. At other sites, we've been able to manufacture products closer to the target specifications. Making cheese is a biological process. Thanks to analytical tools connected to the MES infrastructure, we can now take measurements and use the measurement data just an hour later.

Traceability has not been a primary goal for our projects, but we get it more or less for free with the MES. We had already set up an ISA-95 model of our plant, anyway. It's not a big add-on to achieve full traceability at the same time. We don't use these data so much for recalls (we hardly have any). We use them more to figure out in daily operation where the losses are, and the optimization opportunities. For example: it can make you aware that you always lose two hundred liters of milk in the cleaning process. So always try to focus on the optimization of the process, and then you can see the money that you'll save.

3.2 The MESA Report "Metrics that Matter"

Fortunately, market research agencies haven't let us down. For example, at its members' request, MESA commissioned a study into the way in which manufacturing companies improve their financial performance, and how they justify their investments into production automation software. MESA hired the consulting firm Industry Directions to create an analysis program. The analysis team's Internet questionnaire produced 151 valid responses.

The group of respondents comprised a reasonably even mix of end users from various industries, of which some produce in batches and others use continuous, discrete, or hybrid processes. The results from the study, which were published in October 2006, are thus applicable to industry in the broadest sense of the word. The groups were too small per industry type to draw any meaningful conclusions at the subsector level, according to the report.

One of the questions on the questionnaire was which operational and financial results companies measure. There are probably just as many results that get measured as there are manufacturing companies. That's why the questionnaire concentrated on nineteen familiar operations KPIs (**Key Performance Indicators**) and nine fairly widely used business KPIs (see figures 3.1 and 3.2).

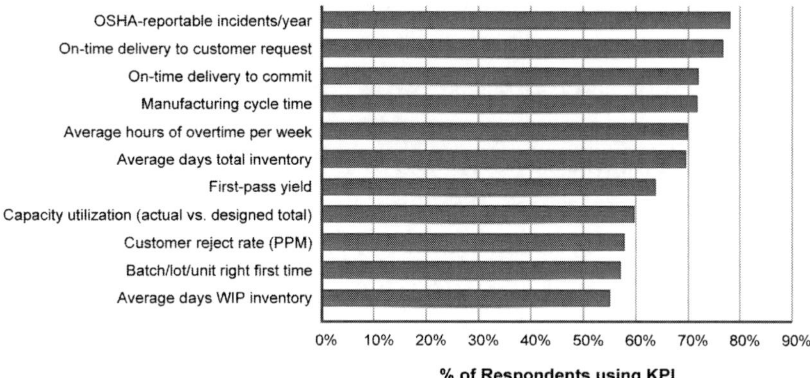

Figure 3.1 Operations KPIs Used by Most Manufacturers, from "MESA Metrics that Matter: Uncovering KPIs that Justify Operational Improvements," © 2006 MESA International & Industry Directions Inc.

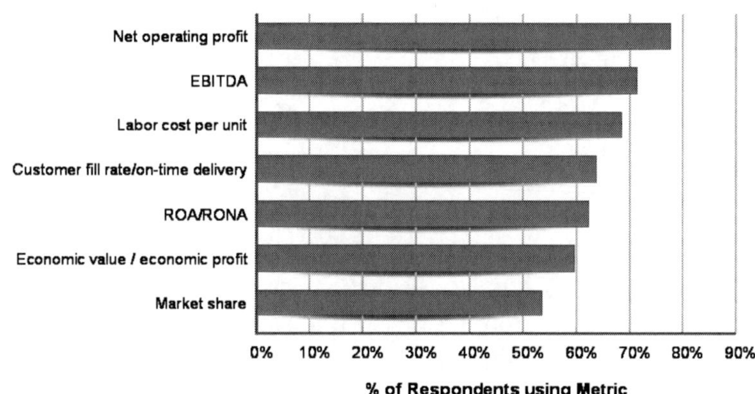

Figure 3.2 Most Manufacturers Use These Business Metrics, from "MESA Metrics that Matter: Uncovering KPIs that Justify Operational Improvements," © 2006 MESA International & Industry Directions Inc.

Of the nineteen operational KPIs, eleven were in use by more than half the respondents, according to the report. The most commonly cited KPI concerned OSHA (Occupational Safety and Health

Administration), the organization to which U.S. companies must report accidents and safety incidents. Two of the frequently cited KPIs related to timely delivery.

Turnaround time turned out to be another important criterion. This indicates the degree to which the plant can quickly adapt to changes. To realize a quick turnaround time, it's important that companies not only design their processes well, but also schedule well, ensure high quality, maximize production results, minimize changeovers and down time, and ensure that materials keep moving along through the process. Because so many issues influence the turnaround time, it's an excellent criterion for measuring plant performance, according to the report.

Other frequently used KPIs concerned inventory size, overtime, and quality.

The report further stated that operational goals are often in conflict with financial and business goals. Julie Fraser, Principal Project Lead Analyst at Industry Directions, gave me the following example in an interview at that time.[2]

> Many companies today have a business goal to innovate, or to increase opportunities within their product lines, and that's why they offer more product varieties. This is at cross purposes with the plant's results and efficiency measurements, since this greater product variety leads to more frequent changeovers, and to operator tasks that are more complex. For example, they have to select the right materials and input the right parameters.

2 Bianca Scholten, "Waarom Investeren in MES en Plant Dashboards? MESA publiceert rapport 'Metrics That Matter'," *Automatie* (no. 3, 2007), 4.

Based on the KPIs listed, respondents to the MESA study indicated how many improvements they had realized in the past three years. The consulting team then divided the companies into two groups, Business Movers and Others. The **Business Movers** are those companies that demonstrated considerable improvements, in breadth or in depth. That is, they started performing more than 1 percent better on six of the eleven business metrics in the study, or they demonstrated more than 10 percent improvement on at least one of the business metrics. These Business Movers serve as a model for other companies. What qualities did they have that were possibly the source for realizing so many improvements?

The report describes the profile for these companies as follows: They're fast; they've coupled their operational goals to their financial and business goals; they know their results; their plant activities are profitable; they concentrate on what's important; they use software applications; and they have, in general, an ROI of two years or less on their investments in plant software.

Figure 3.3 shows the primary areas in which the respondents improved their performance.

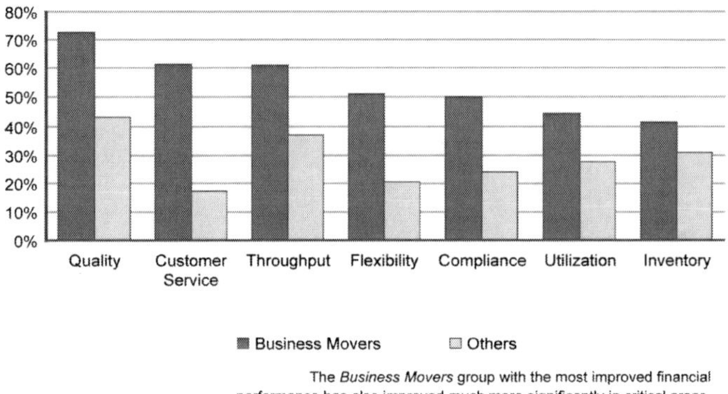

Figure 3.3 Areas of Outstanding or Medium Improvement, from "MESA Metrics that Matter: Uncovering KPIs that Justify Operational Improvements," © 2006 MESA International & Industry Directions Inc.

I asked Julie Fraser to what degree we can know with certainty that the use of software is one of the factors contributing to these positive results.[3]

> We can't know for sure. And in the report, we indicate that there may not be a causal link. All we can say is that those companies have the stated qualities, and that they showed more improvements in their business performance. But considering the number of qualities these companies share, it's unlikely that this is pure coincidence.

According to the report, speed is one of the characteristics of the Business Movers. The faster the company can feed results back to employees, the faster employees can take corrective action. Automated data collection can help make this information available earlier. Many Business Movers feed results back within twenty-four hours, or even in real time. They more frequently use automated data collection than do others.

Other striking conclusions from the report are that Business Movers more often use MES and dashboards (see figure 3.4), and that MES and dashboard users clearly show more improvements than do companies that don't use those kinds of systems (see figures 3.5 and 3.6).

3 Scholten, "Waarom Investeren," 4 (see note 1).

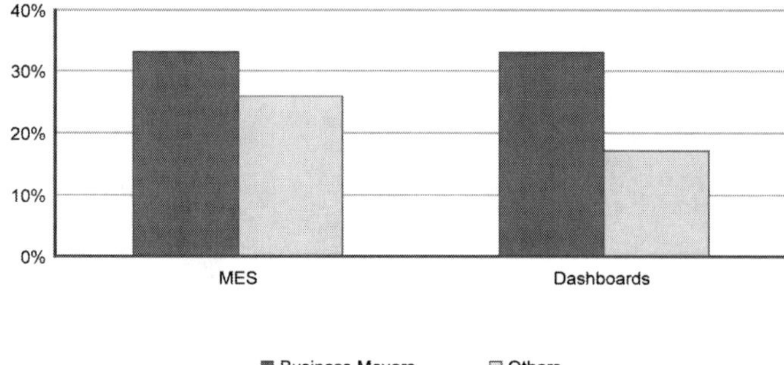

Figure 3.4 Use of MES & Operational Dashboards, from "MESA Metrics that Matter: Uncovering KPIs that Justify Operational Improvements," © 2006 MESA International & Industry Directions Inc.

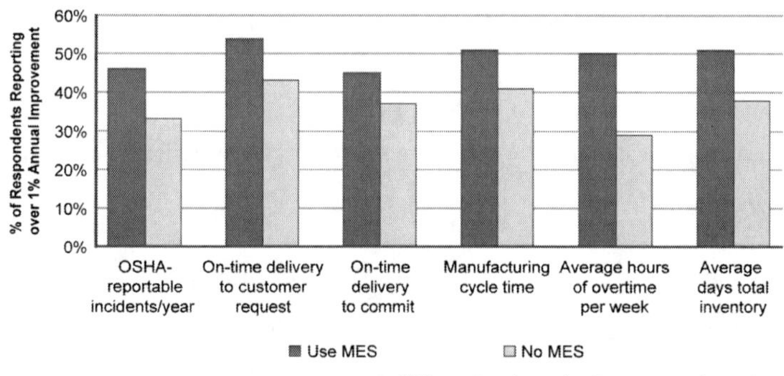

Figure 3.5 A Larger Portion of Those Using MES Improved at Least 1% Per Year on Top Operations KPIs, from "MESA Metrics that Matter: Uncovering KPIs that Justify Operational Improvements," © 2006 MESA International & Industry Directions Inc.

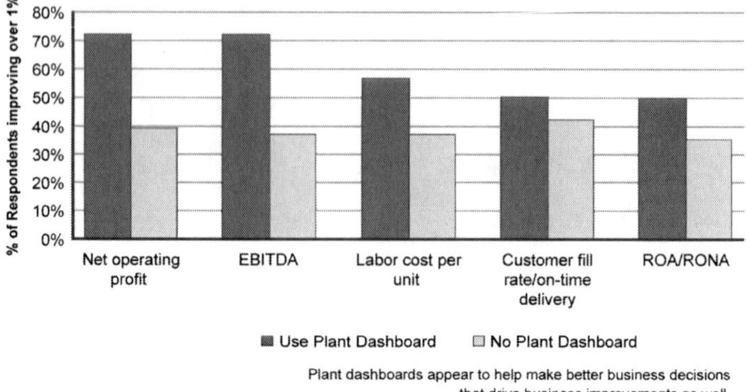

Figure 3.6 More of Those Using Plant Dashboards Improved over 1% Annually on Most Used Financial Metrics, from "MESA Metrics that Matter: Uncovering KPIs that Justify Operational Improvements,"
© 2006 MESA International & Industry Directions Inc.

Many respondents couldn't provide information on the ROI of their current systems, states the report. The question arises, why should one take the trouble to calculate it? After all, the system has already been implemented, the money's spent, and you can't undo any of that. Julie Fraser explains.[4]

> It turns out that companies do indeed feel this way. The most important reason to measure ROI is to prove to skeptics and to top management that the investments they've already made have been advantageous, and to increase their enthusiasm for future investments. One of the greatest problems companies face today is that their financial systems aren't set up to register the actual profit realized through improvement projects. In other words, companies don't run their plants based on a realistic view of what does and doesn't work.

4 Scholten, "Waarom Investeren," 4 (see note 1).

> The most important reason to measure systems' ROI may well be to know for certain that you're taking full advantage of the plant's possibilities, and that it gives you the opportunity to minimize outsourcing or offshoring. Many manufacturing companies close plants in areas with high production costs, such as the US and Europe. The people who work in those plants will lose their jobs if they can't prove their efforts contribute to reaching the company's corporate and financial objectives.

The report "Metrics that Matter" contains many more interesting insights. You can find it at www.mesa.org.

3.3 Advantages per Functional Area

MES is a belief, but various market research agencies think they've found a causal link between the use of MES functionalities and improvements to operational and company-wide performance. Keep in mind, however, that these studies aren't always conducted at the request of an independent organization, but often on behalf of vendors. In that case, you might question the market research agency's objectivity.

Ultimately, of course, we have to keep using our own sound judgement under all circumstances. Let's try to explain those expected improvements just by thinking logically, assuming the manufacturing execution system functionalities defined in chapter 2: detailed production scheduling, product definition management and production execution management, historians, reporting and tracking & tracing modules, dashboards, workflow management, integration with ERP systems, and plant models.

3.3.1 Detailed Production Scheduling

Using a detailed production scheduling package may possibly lead to higher efficiency. Such a system can quickly calculate diverse options (simulation) and, in so doing, propose the most efficient combination and sequencing of orders, changeovers, and cleanings. In companies where there are many disruptions or rush orders, for example, a detailed scheduling system can help to determine the impact the changes will have, and to adapt the schedule.

For a scheduler who has to do all this manually or using Excel, it's virtually impossible to spend time reviewing all the different possible schedules. It requires far too much calculation. A system for detailed scheduling calculates appreciably faster, and can thus give insight into how you can optimally use capacity. It's possible that, thanks to the scheduling package, you'll produce more in the same amount of time, or avoid overtime or having to invest in extra capacity.

Some detailed scheduling solutions can also provide insight into short-term inventories, and thus support just-in-time deliveries. That might lead to a reduction in idle-time fees the company has to pay for transport. And who knows; maybe safety stock levels can also be decreased. Then the dollars saved are truly measurable.

Consider also increased flexibility, which allows you to respond better to individual customer requirements. The scheduling system receives orders and can send the customer's specific wishes along with the order to the shop floor, which potentially leads to improved customer service.

If the detailed production scheduling system is given an automated interface to the Level 4 scheduling system, the scheduling information will be in the right place faster. You won't need as much phoning back and forth, and you reduce the risk that people will be working with an outdated schedule.

3.3.1.1 From the Trenches: Advantages of the Scheduling System at Masterfoods

In June 2005, I interviewed Gerard Alders, IT Manager for Masterfoods' European plants, about Masterfoods' new detailed scheduling system.[5]

> On Friday, the line schedule is ready and the program reads it in. Then we know the exact amount to be consumed. Often, there are small differences from the previous plan, and the chocolate room adjusts the current plan by (for example) having a truck arrive a little earlier or later.
>
> They may be small differences, but nonetheless the system delivers a significant reduction in costs. The first benefits were visible as soon as we started using the system. Better insight into the exact moment when certain raw materials are needed has led to a significant reduction in truck idle times, and with it a savings in idle-time fees.

3.3.2 Product Definition Management and Production Execution Management

If you implement a module for recipe management, assembly instructions, or SOPs, then you can synchronize master data automatically with the BOM in ERP. Particularly for companies in which recipes often change, this delivers advantages. There's a smaller risk that operators will be using outdated instructions, and changes from R&D can be more quickly incorporated. Because the recipes are flexible, you can always use the least expensive raw materials, as is common practice in the feed industry.

5 Bianca Scholten, "Scheduling op basis van S95. Gedetailleerde planning van chocoladeproductie bij Masterfoods," Automatie (no. 7, 2005), 10.

The software can also automatically send product-related parameters to the PLC-SCADA layer. Because operators no longer have to manually transfer those data, they win back time, and the risk of error is reduced. Now, instead of registering data, they can focus on monitoring and analyzing the process.

In pharmaceutical companies, where one operator sets the parameters and a second must check them and confirm that the first operator has input them correctly, automating this process might even lead to a reduction in FTEs. Thanks to a validated, 21 CFR 11-compliant[6] interface, you no longer need operators for the exchange of information.

Remote control also becomes possible. As a result, the operator is less frequently exposed to dangerous situations, and the product runs less risk of coming into contact with contaminating agents.

But even if the MES doesn't automatically send the parameters to Level 2, the system can still deliver a variety of advantages. Paper instructions can't check and correct the operator. In contrast, an MES can warn the operator at the moment he or she enters data that fall outside the specified limits. So quality could improve, and that could lead to less waste.

This kind of functionality can also give insight into the current status of an order, and help the production manager to focus on orders in danger of missing their deadlines and orders having the highest priority.

6 21 Code of Federal Regulations Part 11 concerns electronic records and signatures.

3.3.2.1 From the Trenches: Integration of Product Definition Information at International Flavors and Fragrances

In November 2005, I spoke with Jos Jaspers, Business Engineer at IFF's SAP Solution Center, about the benefits of integrated information.[7]

> Integration produces important benefits for IFF in terms of error sensitivity and the speed of information exchange. And integration also allows us to achieve more consistent quality. People, who could make the wrong decisions, are no longer involved. Machines don't do that. They're more precise.
>
> It's very important that we deliver consistent quality. Together with the customer, we develop a product line for (say) perfume, soap, and shampoo. The customer expects the quality of the fragrances to remain the same for as long as they use them in their products.

3.3.3 Historians

A historian can take over the operators' task of collecting various data, so they can concentrate on their real task: monitoring and guiding the process. Moreover, a historian gives you the opportunity to store large quantities of data over a long period of time in a central location. This improves access to historical data and lays the foundation for process analysis and optimization.

3.3.4 Reporting and Tracking & Tracing Modules

In many plants, people use paper forms and Excel to create reports.

7 Bianca Scholten, "Van SAP tot PLC; Hoe bereik je 100% integratie?," *Automatie* (no. 1, 2006), 4.

For example, there are pharmaceutical companies that produce more pounds of paper batch reports than they do pounds of end product!

Customers and government require tracking and tracing, and as long as you're recording data, why not use all that information for other objectives, such as continuous improvement? But collecting data from paper forms and Excel files, converting units of measurement, and then making comparisons costs process engineers and production managers an awful lot of time—time they'd rather spend improving the process, quality, efficiency, and so on. Moreover, it's virtually impossible, using those separate sources, to discover (for example) how a certain parameter in the raw material preparation process causes an error in the finishing process. So these people suffer a serious lack of insight.

MES can provide that insight, by quickly generating a variety of reports and charts based on the collected data, relating to quality, OEE, performance by shift, by day, by batch, and so on. These reports form an important input for team discussion, in order to get everyone on the same wavelength and to concentrate on what's really important.

It's impossible to calculate beforehand the improvements to which all this ultimately leads. Who can say? Maybe quality will increase, or waste will decrease. Maybe you'll be able to avoid overtime, shorten turnaround time, lower safety stocks, produce more in less time, and a hundred other things. You won't know until you have accurate insight into the current situation.

3.3.5 Dashboards

But reports alone won't get you there, unfortunately. They do offer a basis for making various comparisons, but by definition they look at the past, and they always contain the same information. Reports are static.

A company that runs its plant based on reports is like a car owner who drives while looking in the rearview mirror. It's better to know what's happening now. To respond proactively. To make adjustments. And because you can't be everywhere at once or talk to everyone at once, it's useful to have a dashboard.

A car dashboard tells you whether you're driving too fast, what your oil level is, whether the temperature is okay, and whether you're going to arrive at your destination on time. If you're smart, you fill up your oil when it's convenient instead of waiting for the lamp to come on, which, of course, always happens when you're driving to an important appointment. If the road is blocked or there's a traffic jam, you take a detour. If the system detects a speed trap, you step on the brakes.

In the same way, you can use a plant dashboard to view various aspects of the production process. The advantages are evident: you reduce turnaround time, your customers are happy, you limit maintenance costs, you avoid government fines, and so on.

3.3.6 Workflow Management

There's a good chance that the production department uses a variety of information systems, and that logistics, the lab, the warehouse, and the maintenance department all have their own, separate systems.

"It's like an orchestra," says Henk de Man, Senior Architect at Cordys.[8] "Every system plays its own instrument. One plays guitar, the other piano. But the music they play together is constantly changing, so they have to continuously adapt and attune their playing to the new music and to each other—a difficult task at best."

8 Bianca Scholten, "Productie als schakel in de keten; Business Process Management Systemen dirigeren het orkest," *Automatie* (no. 4, 2008), 20.

How do you avoid a cacophony of departmental melodies out of sync with one another, so that the result makes you want to cover your ears?

I know a few examples of companies where employees don't bother signing orders in and out, so that no one knows which process step the order is in. If part of the order is rejected, the order isn't closed out, which leads to a financial drama. Months later, still no invoice has been sent. And then you've got office assistants who keep calling the warehouse to ask whether a shipment is finally ready for transport. Time after time, they shuffle through the papers to check whether everything's finally, really complete. Lean manufacturing, you were saying? They waste enormous amounts of time repeatedly starting activities, only to be stopped in their tracks because colleagues haven't yet provided missing information.

Workflow management systems can function as the orchestra conductor. They streamline processes in which different departments are involved, by pointing out tasks and priorities to employees and by providing insight into who's working on which order, and where. Implementing a workflow management system could thus lead to shorter turnaround times and improved delivery reliability, in particular.

3.3.7 Interfaces and Integration with ERP

If the MES platform has an automated interface to Level 4, then your ERP system will in theory have more current and more correct data available. This lowers the risk that, for example, inventory data are erroneous, which in the worst case could lead to production stops.

More current and realistic inventory information on Level 4 offers you the possibility of lowering safety stocks. Moreover, an interface like this can help supply the ERP system with information on the actual production hours per product, the personnel used, and other information you can use to calculate the cost price. That leads to better insight into which products generate the highest margins, and which products you're better off removing from your product line, because there's hardly any difference between the cost price and the retail price.

3.3.7.1 From the Trenches: Integration with ERP at Akzo Nobel

Akzo Nobel was confronted with several challenges when they established an interface between SAP R/3 and Proficy from GE Fanuc at their site in Herkenbosch, the Netherlands. Ruud Swarts, Process Engineer and project member on the MES and integration projects, Wouter Amersfoort, Information Manager for Logistics at Akzo Nobel, and Nick van der Schilden, consultant at Ordina TA[9], explain.[10]

> **RS:** In Herkenbosch, we manufacture the raw materials for various industries, such as the food industry, pharmaceuticals, agriculture, and gardening. We have about fifteen processes, most of them batch, and a few semi-continuous. Herkenbosch uses more than 150 recipes and is in production twenty-four hours a day, seven days a week. Even though it's not a very large plant, there's a lot of diversity, and the processes are relatively complex. As a result, the process and product data are also complex, and it's difficult to extract key performance indicators from them.
>
> In the past, people had to enter the same data in up to four different recording systems and multiple fields, which generated a lot of errors. And it cost a lot of time. It was more or less a full-time job to check and correct the data entered, to

9 TASK[24] since 2008.

10 Jan van Bekkem, "Zegt het ene systeem tegen het andere; Akzo Nobel integreert SAP en Proficy," *Automatie* (no. 6, 2007), 10.

fill out forms completely; in short, to make sure everything was accurate in the books.

WA: We haven't completed the full rollout, but the first results are impressive. Thousands of messages are being sent. Right now that only happens in part of the plant; later it will happen in the whole plant. It's resulting in an enormous time savings and in improved quality in the data exchanged.

We've had a study done, and it showed there's a 10 percent error rate in manual data exchange. Another benefit is that we now have the actual consumption numbers available in the ERP. Over the long term, those will probably lead to more insight into our inventories. In the past, people entered the [production] results into the ERP as they were supposed to, but not the amounts consumed, because that was too much work for the operator. The result was that at the end of the month or the year, we were constantly busy making correction entries in the ERP. We didn't have a good sense of our inventories.

NvdS: Master data management is also much more efficiently set up now. In the past, Akzo Nobel had to maintain the bill of materials in two places. Now, however, the ISA-95 interface sends a copy of the BOM with the control recipe, so the MES always receives the right BOM. If the BOM changes in the ERP, then it automatically changes in the MES, too.

3.3.8 Plant Model

At the heart of every MES solution is a **plant model**: a configuration of the production processes, products, materials, and so forth. For example, a dedicated scheduling solution contains such a plant model. If you buy a recipe management system from a different vendor, it will contain its own plant model. And an OEE module from yet another provider can't live without this functionality, either; it will contain a third plant model. If the MES landscape within a manufacturing company comprises "best of breed" solutions from diverse vendors, you run the risk that these systems will be configured in significantly different ways, and that it will thus be difficult to keep information in the different systems synchronized.

Synchronizing and managing the master data in these systems is a time-consuming and error-prone process. MES vendors realize this, and slowly but surely they're offering databases and modeling functionality that conform to the ISA-95 series of standards, so that the total MES landscape can build on a single, unified view of the processes and their associated information. This leads to several advantages for end users. For example, you need less time to synchronize systems. The whole system also becomes less complex, and the risk of error decreases.

In short, by thinking logically, it's fairly easy to imagine all the advantages a manufacturing execution system could generate. It goes without saying that the most important opportunities vary from company type to company type. A company that continuously produces the same product probably won't gain much from a scheduling package and a recipe management system. For companies that use the same machines to make a large number of different products, a detailed production scheduling solution is probably relevant. Different industries place emphasis on different things, and every company should weigh the costs against the benefits for itself.

Speaking of costs: what does an MES cost, actually?

3.4 Costs

What does an MES cost? That question is just as hard to answer as "What does a car cost?" The costs for an MES are determined by many factors:

- The number of plants and the number of production lines within the scope of the project.
- The opportunities for reuse of functionality if, for example, production lines or reports closely resemble one another.
- How extensive is the requested functionality? Just an OEE module? Or also a historian, a recipe management module, a detailed scheduling solution, ...?
- How many interfaces to other systems (such as LIMS, ERP, and PLCs) need to be realized?
- How many licenses do you need? How many users will there be?
- Do you need to purchase hardware (such as workstations, sensors, PLCs, servers), or are these already available?
- Does the PLC software need to be altered?
- Rates, price differences among vendors (not only the software vendor, but also the system integrator)
- Are specification phases (user specification, functional specification) also part of the proposal?
- Will project management from the vendor also be included in the proposal?
- Is project management on the client's end (implementation guidance, user training, coaching, and so on) included?
- And so forth.

In short, when vendors give an indication of the investment required for MES, don't look at the price alone. Many vendors glide in like the Trojan horse: "Our solution only costs thirty thousand dollars!" Once you've signed on the dotted line, the truth comes out, and you discover that essential items haven't been included in the price. So compare apples and apples!

But broadly speaking, what should you consider? After all, you can list some general guidelines for buying a car. What about for an MES? Will it cost a few ten thousand dollars? A few hundred thousand? Millions?

During MESA's European conference in 2007, Reinoud Visser, Principal Consultant at Atos Origin, presented the results of a survey, including several interesting experiential numbers (see figures 3.7 through 3.11).[11] The respondents, in total 395, included not only manufacturing companies, but also system integrators, consultants, and vendors. The study investigated MES projects carried out from September 2006 through September 2007. The charts speak for themselves.

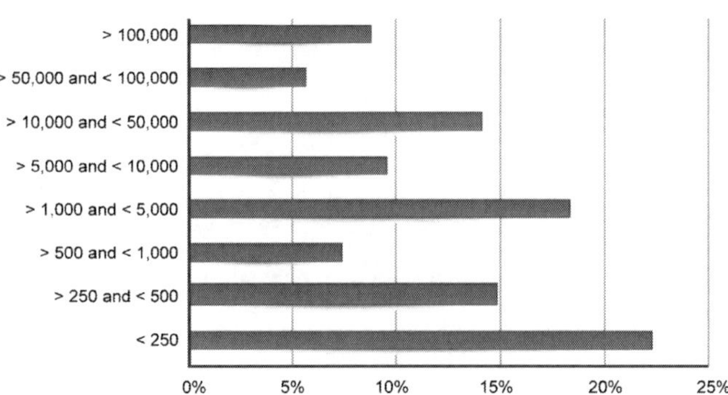

Figure 3.7 Company Size (Source: MES Experience Survey, Reinoud Visser, Atos Origin, presented during the MESA European Conference, 2007)

11 Reinoud Visser, "MES Experience Survey" (presentation, MESA European Conference 2007, Utrecht, November 2007).

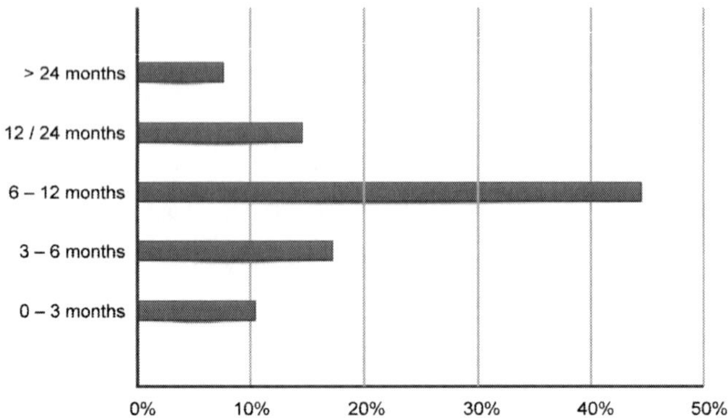

Figure 3.8 Implementation Time (Source: MES Experience Survey, Reinoud Visser, Atos Origin, presented during the MESA European Conference, 2007)

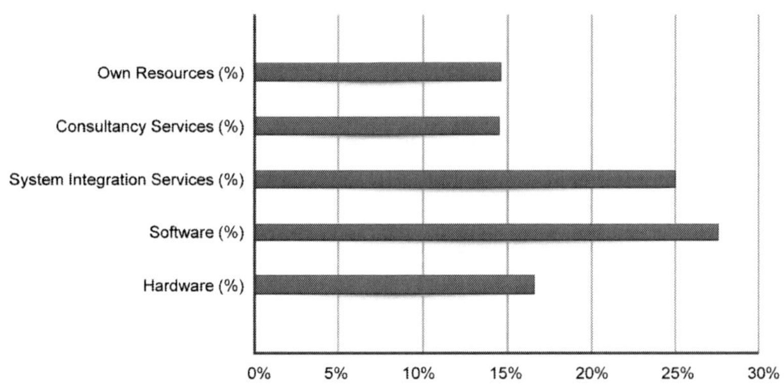

Figure 3.9 Budget Resources (Source: MES Experience Survey, Reinoud Visser, Atos Origin, presented during the MESA European Conference, 2007)

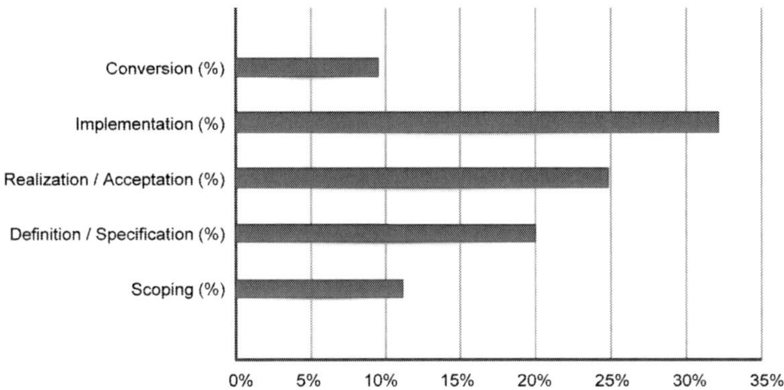

Figure 3.10 Budget Phasing (Source: MES Experience Survey, Reinoud Visser, Atos Origin, presented during the MESA European Conference, 2007)

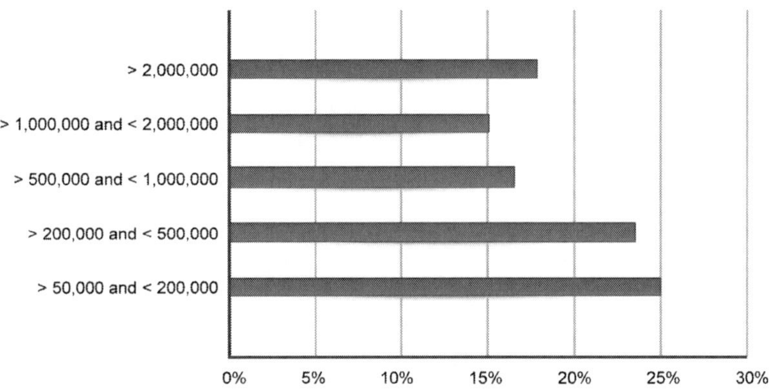

Figure 3.11 Budget Size (Source: MES Experience Survey, Reinoud Visser, Atos Origin, presented during the MESA European Conference, 2007)

Okay. Now we have a general idea what it would cost, and an idea of the benefits it can provide. To get an even better idea, you could request a few quotes from different vendors yourself, of course. But to do that, you first have to have a common, clear idea internally of what you all want. And then you have to select a few vendors. Who should you involve in the internal decision-making? How do you then prevent the decision-making process from becoming an infinite loop? And where do you find the MES vendors that offer suitable solutions? Chapter 4 answers these questions.

CHAPTER **4**

How Do You Buy an MES?

If it looks like an MES is something your company can use, then it's time to carefully chart your needs and desires, go looking for a suitable solution, and choose the party who'll implement it. Then you'll need to make it clear to the vendor what you expect from her. What's the best way to do that? Who's going to write the requirements, and what exactly should be in them? Where do you find an overview of MES vendors? And what are the steps in the selection process? In other words: how can you quickly and efficiently purchase an MES that meets your company's needs as fully as possible?

4.1 It's Not Easy Being a Client

I bought a house last year. It was perfect, except for the bathroom. That summer, I visited the showroom of a well-known bathroom supplier.

"Would you like some coffee?" the woman behind the desk asked me.

"Yes, please." I was sure that buying a new bathroom would take a lot of table time with the salesperson, so I could use a cup of coffee. But to my surprise, I ended up all alone with my coffee. This isn't working, I thought. I waited until I saw the woman again, and waved to her.

"I'd like to talk to someone about bathrooms."

A few moments later, a young man came toward me. I introduced myself and said, "I want to buy a new bathroom."

He stared at me for a moment before asking, "And you'd like to make an appointment to discuss that?"

"No. I'd like to discuss it right now!"

"Oh, well…we can do that…please take a seat while I get my materials."

He came back with some paper, a box of colored pencils, and a measuring stick. I told him the size of my bathroom, the locations of the window and the door, and that I wanted a bathtub, a shower, a toilet, and a sink. He made a few drawings of possible floor plans. After that, we walked through the showroom and I pointed out the bathtub, shower, and all the other things I wanted to buy.

"When will the price quote be ready?" I asked.

"You'll receive it within a week, by regular mail."

A week later, I did indeed receive an envelope. The salesman had enclosed the floor plan and a price: eight thousand dollars. There was nothing else in the envelope. I decided to call him.

"Thank you for the price quote, which I've just received. But you forgot to enclose the specification list. And moreover, it doesn't say whether the price includes installation."

"It's always without installation. We never install the bathrooms ourselves."

"Okay. But could you please send me the list of specifications for the items I selected?"

"We never do that."

"But then how can I be sure you're going to deliver the things that I chose? And what if I think it's too expensive? How can I know what the most expensive items are, so I can choose cheaper ones?"

As you've probably guessed, we didn't reach an agreement. The sales guy just kept on saying that this was their usual way of working. So I said goodbye and hung up. *I'm not going to buy a bathroom from them!*

This bathroom story has become an oft-used metaphor in my office for poor communication between customer and vendor. How do you choose a suitable MES vendor and a solution that fits your company? And how do you communicate clearly and unambiguously with that vendor about your company's MES needs and desires?

4.2 Determine the Scope

MES is a broad concept, and it comes with a price tag. Just as for buying a bathroom, it's important to know what you do and don't want in an MES. That's step one: determining the scope. The ISA-95 models can be useful tools for that (see figure 4.1).

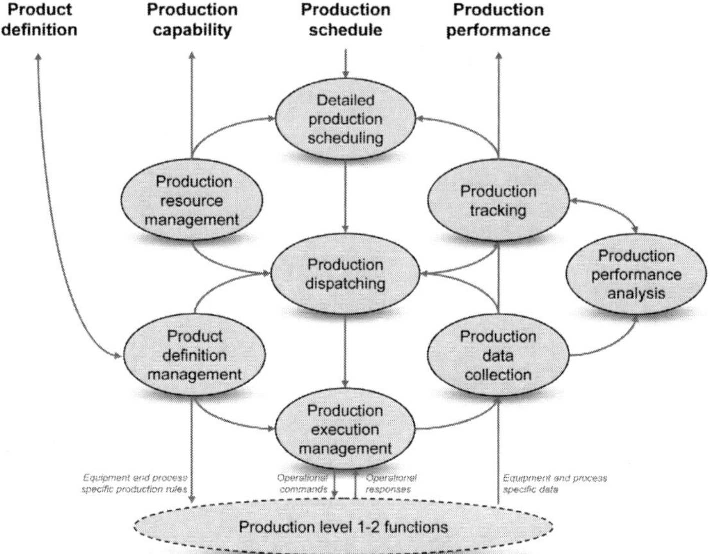

Figure 4.1 ISA-95 models, such as this Production Operations Management model, can help you determine the scope of your MES project

Do you limit your scope to data collection, for example, or do you also include detailed production scheduling, product definition management, and so on? Note that even if you don't want to invest in that other functionality now, it can still be important when choosing a vendor to know whether they'll also be able to meet your future requirements. That way, you can avoid having your MES landscape slowly but surely turn into a patchwork of different packages, which leads to disadvantages such as having to maintain interfaces, dependence on knowledge of diverse technologies, and unnecessary complexity in managing master data.

On the other hand, it can be a waste of your time to pore over future needs now. A lot can happen that would make it all wasted effort. The most logical approach, of course, is to define the short-term requirements in detail, and only map out your longer-term needs at a high level.

How do you do that, map out MES user requirements?

4.3 The User Requirements Specification

4.3.1 Purpose of the URS

During a forum discussion, an IT manager at a chemical plant said, "Those fat documents that consultants write? Nobody reads those."

A feeling of "You talkin' to *me*?" crept over me. As an MES and vertical integration specialist, I churn out those "fat documents" regularly. I speak with the board, the managers, and the users of the new MES system to be purchased. In interviews, they tell me what they expect from the new system, and how it should provide advantages. I turn these findings into a **user requirements specification** that easily contains a hundred pages.

For many end users, maybe a document like this does end up at the bottom of a drawer somewhere. But my colleagues, who have to build the system, use it repeatedly, and they're always very happy to have it. The better the client's processes and needs are described, the easier it is for my colleagues to ensure the system meets them. Ultimately, it increases the likelihood that they'll deliver a suitable system, and that the client will be satisfied.

That IT manager's comment has prompted me to explicitly describe to clients, at the start of a URS trajectory, why the document is important. "You all might not use it," I say, "but the vendor has to have an extremely good understanding of what's involved here. That's why I describe the unique characteristics of your processes. In selecting a package, we can then check whether the solution is a good fit for your company. And then the system integrator will use the document as a basis for realizing the solution."

I also gave this explanation at a research company where I conducted an extensive study into the desired MES functionality in various departments. The users of the future system were highly educated people, so it was particularly important, in order to have them accept the new MES, that everyone could put in his two cents' worth. When the document was finally finished and accepted, we were able to start selecting a vendor.

"So," I commented in a conversation with the internal project leader, "now the document's going to prove its worth, because the vendors will be getting solid information about your processes and desires and requirements, and they can make a proposal that meets them."

"Yeah," he answered, "you keep saying that the URS is particularly useful to the vendor that's going to provide the solution. But in my experience, it's also an important document for us internally. We didn't go through the whole specification trajectory that we just completed purely to provide the future vendor with information. The project has also particularly helped us to come to internal agreement about what we want. As a result of all the interviews and workshops you conducted, we've gained clarity on all the functionality the system needs to have. And on what the fundamental purpose of the system is."

This client opened my eyes. The URS isn't just a document that specifies the user's processes and desires and requirements so that the vendor knows what's expected of him. The URS also records the decisions to which the internal parties have jointly agreed. The URS is final once the stakeholders have signed their names to it.

4.3.2 Contents of the URS

So one of the purposes of the URS is to provide the vendor with sufficiently detailed information about your company, and about your desires and needs for the MES. But exactly what information does a vendor need in order to make a proposal?

When I first started writing user requirements for MES, I used the ISA-95 series as a guideline. ISA-95 did turn out to be a useful starting point, but the scope of the standard is broader than MES, and conversely, a URS should contain more information than just the subjects covered by ISA-95. Within the ISA-95 & MES Competence Center at TASK[24], we decided to conduct a study on best practices for writing a URS for an MES.

Our analysis showed that most authors of URS documents seem to have a clear picture of its purpose. Nevertheless, when comparing URS documents written by different authors (even authors within one and the same company), there appear to be huge differences in scope, content, and level of detail. We decided to develop a URS template based on best practices, so we'd have a guideline to help our customers write high-quality URS documents. The template we developed is also a big help for MES vendors' sales teams. Using it as a checklist, they're able to decide whether the information the customer has provided is complete and of sufficient quality to be able to make a proposal.

Appendix B contains the table of contents for this template.

4.3.3 Gathering Requirements

The template referenced in appendix B is a good guideline for finding the people who can provide the necessary input. Gather this information by organizing internal workshops and interviews, and by consulting existing documents. In my book *The Road to Integration: A Guide to Applying the ISA-95 Standard in Manufacturing,*[1] I describe in chapter 2 how to approach such an analysis. The chapter also contains several tips and points of concern.

In any case, set up a team, and make a project plan. If desired, you can hire an external consultant to lead the workshops and interviews, and to write the documents. That does cost more, but it also provides important advantages.

To begin with, it reduces the workload for the company's employees, who would otherwise have to do this in addition to their existing tasks. It also creates more pressure to meet the deadlines. After all, it's harder to push back the date on agreements you've made with an external party than on agreements you've made with colleagues.

The external consultant can also help you focus on the main issues and avoid being sidetracked into all kinds of irrelevant internal discussions. That's a tendency that rears its head all too soon when you gather employees from different departments around the table.

1 Bianca Scholten, *The Road to Integration: A Guide to Applying the ISA-95 Standard in Manufacturing* (Research Triangle Park: ISA, 2007).

Moreover, people can be more open and honest with an external consultant about certain problems, without worrying that they'll step on someone's toes. Finally, there are many consultants who specialize in generating a URS for MES projects, while internal employees would probably be doing it for the first time.

4.3.4 Writing Requirements

The URS for an MES contains not only a lot of textual and graphical background information, but also an appendix with a numbered list of requirements. This list is a special point of concern. It's a tool for the end user to check whether the solution fulfills all the user requirements. It's also the basis the vendor uses for making a cross-reference table that shows which part of the solution meets which user requirement. In order to avoid misinterpretations, it's important that the requirements are defined in a **SMART** way, meaning **S**pecific, **M**easurable, **A**cceptable, **R**ealistic, and **T**ime-bound. That's easier said than done, however.

In 2008, I interviewed Wim Dijkgraaf, one of the authors of *Start at the End with SMART Requirements,*[2] for the trade magazine *Automatie.*[3] The book explains how to write requirements properly. Using an imperfect example as a starting point, the authors apply guidelines for describing the requirements. For example, you should describe the need and not the solution. *The coffee machine has a timer* isn't a requirement, because it represents a solution. *The user is able to plan the availability of coffee* is a requirement, because it posits a need.

In the book, the authors also discuss how to validate the result. Wim Dijkgraaf explains.

2 Wim Dijkgraaf and Mike van Spall, Start at the End with SMART Requirements (Chaam, the Netherlands: Synergio BV). The English version of this book is available under ISBN 978-90-78990-02-4.
3 Bianca Scholten, "Wat wil je, wat krijg je; Het schrijven van Smart Requirements," *Automatie* (no. 7 and no. 8, 2008), 14.

If you want to determine whether the solution is satisfactory, you need to have made your requirements measurable. You have to distinguish between a quantified requirement (a number and a unit) and a measurable requirement. You make a requirement measurable by assigning it one or more indicators. An example will help clarify that: The coffee should be warm enough. *Warm enough* is vague, so you want to quantify it: 115°F (46°C), plus or minus 5°F (3°C). Now it's quantified, but not yet measurable. You still have to define an indicator, such as *percentage of coffee temperature readings within the norm*, for which the target value is, say, at least 80 percent. This means that in at least 80 percent of the cases, the coffee's temperature must meet the stated target temperature.

Measurable requirements are important for testing (verifying, validating). This is the perfect input for a testing department, which then determines how it can measure the indicators in a cost-effective and sufficiently reliable way.

No matter how clearly you've worded the URS, if people haven't been part of the thought process, it's hard for them to read it exactly the way the person who wrote it does. So it's very important to take part in the document's creation. Reading it on paper doesn't give a person half the understanding that people who've participated in the process have.

So remember that a requirements document is the *beginning* of communication between the customer and the vendor, and not the end. Be especially sure to keep communicating verbally over the course of the project. Never expect that you can just toss the document over the fence and the vendor will have built a successful solution a few months later!

Use the same structure in the requirements list that you use in the text document. Each requirement gets a unique number and a reference to the section in which the background information is described, as shown in the example in table 4.1.

Table 4.1 Sample possible structure and content for a requirements list

No.	Requirement Description	Document Reference	Priority	Compliant %	Facilitating %	Custom %	Remarks
Production data collection							
22	Description	Sec. 5.6.1	Knockout				
23	Description	Sec. 5.6.1	Nice to have				
Production tracking							
24	Description	Sec. 5.7.1	Medium				

The **Priority** column serves to rank the requirements. Vendors don't get to see whether a requirement is a knockout criterion, a general requirement, or just a nice-to-have, so be sure to delete the column before you send the list to the vendor. It's meant to help you assess the vendors' responses in a neutral manner. Without an independent assessment system,[4] you run a significant risk of selecting the package from the salesman with the best presentation skills, instead of the one from the party who has the most suitable solution.

In the columns **Compliant**, **Facilitating**, and **Custom**, the vendors fill in the percentage that can be provided standard ("out of the box"), the percentage that's configurable by adapting screens, reports, and so on, and the percentage that will have to be custom programmed, respectively. The total of the three must be 100 percent. Depending on whether it concerns a knockout criterion, a general requirement, or a nice-to-have, the score will weigh more heavily or less heavily in the final assessment. In this way, the package is awarded a total number of points, and you can compare this total with the scores for other packages and vendors.

4 The assessment system described here was developed by Peter Bierens, a consultant at TASK[24]'s ISA-95 & MES Competence Center.

Note, however, that the vendors fill in these three columns themselves, so there's a risk they'll paint too rosy a picture of the solution being offered. Stay alert, ask critical questions, and check whether the vendor can deliver on what he's promised during presentations, on any visits you make to references, and in any proofs of concept the vendor provides.

4.3.4.1 From the Trenches: Arla Foods' MES Requirements

Arne Svendsen describes how Arla Foods developed its MES platform.[5]

> We needed a generic requirements specification incorporating ideas from many of our sites. We created one in 2003. You have to be a little bit visionary and try to project what the future requirements from your internal customers will be. And you even have to look at the market and try to understand what kind of functionality will be available in the future.
>
> Of course you want to select a solution that delivers the highest percentage of your requirements in its standard functionality. But whatever solution you select, only a maximum of about 80 percent of the requirements will be fulfilled by the standard solution. So you'll always have to add some modules of your own, or even integrate a module from another MES vendor. That can be hard.
>
> Over the years, we've developed several of our own modules. We couldn't always wait for the release of a first version of some MES functionality from our supplier. We built our own ISA-95 module, in which we created a product definition data model, including our own interface. We integrated this

5 Arne Svendsen (head of Manufacturing Services & Automation, Arla Foods Global IT), interview with the author, September 2008.

interface and the database with the MES platform. This module downloads the latest version of the recipe to the PLCs, in real time. We also built a module for the assessment of cheese. We have a panel with judges who test the cheese. We link the results from these assessments to the batch record in the MES platform.

Some of the modules that we built ourselves will probably be phased out when our supplier comes up with a solution that is at least almost as good as ours, and that's better integrated with the standard MES solution. The fact that we built our own modules doesn't mean we needed a big IT team. They were small modules, and luckily we had a smart guy in the team who built them in just four months, as a .Net application. When you need small modules, it's possible to build them yourself.

So the MES platform needs to have a degree of openness. It should have a set of services with a standard interface between the MES and the module. Your MES solution should have the ability to work with Web services. The concept of Manufacturing-SOA[6] is that within three years' time, you'll have a platform onto which you can add modules from different vendors. The ISA-95 models can be the heart of this platform, because the vendors need one common agreement on what the basic data are for their applications.[7]

4.4 The Selection Process

Okay. The requirements have been mapped out, and the assessment form is ready to be filled in by the vendors. But where do you find

6 Services Oriented Architecture.

7 For more information, see MESA's guidebook entitled "Services Oriented Architecture (SOA) in Manufacturing."

those MES vendors? And how do you walk through the selection process, step by step?

Globally speaking, the selection phase consists of the following steps,[8] which will be explained in more detail in the following sections (see also figure 4.2):

- Put together a long list
- Make a short list
- Send a Request for Information and a questionnaire to the vendors on the short list
- Organize a Q&A session for the software vendors
- Schedule vendor presentations
- If desired, schedule a reference visit and a proof of concept of the solution
- Select the package
- Make a short list of system integrators
- Send an RFQ to the system integrators
- Organize a Q&A session for the system integrators
- Schedule the system integrator presentations
- Select the system integrator and sign the contract

8 This schedule assumes that you first select the product and then the system integrator. That's a matter of preference. One of my clients strongly preferred this approach, given the international character of their project. They knew that no single system integrator could provide support in all the countries involved. That's why they required the product vendor to function as the central point of contact. But you could just as easily select your preferred system integrator first, and let him or her propose a product, or use combinations of preferred MES products and preferred system integrators.

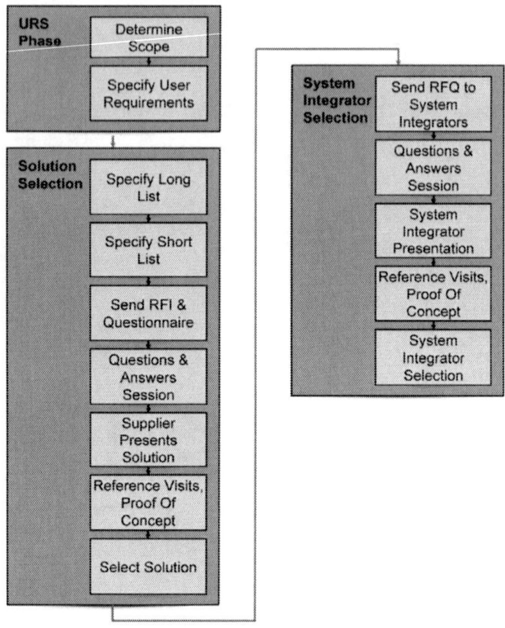

Figure 4.2 A global overview of the steps in selecting an MES solution and a system integrator

4.4.1 The Long List and the Short List

Various sources can provide input for a long list of about six software packages. Perhaps vendors have already visited your company, or you've met promising candidates during shows or conferences. Consult trade magazines; these sometimes write about MES projects and solutions. Maybe other plants in your division already use particular MES solutions that might be suitable for this plant, too. Another useful tool is the MES Product Survey that Logica publishes annually. You can find it on the MESA website, among other places. This document provides an overview of dozens of MES vendors worldwide, and a high-level description of their packages.

To assess whether a vendor or package belongs on the long list, you define several high-level knockout criteria. For example,

"The package is suitable for both batch processes and continuous processes," or "The vendor provides support throughout Europe," or "The package contains detailed scheduling functionality." Ultimately, this results in a list of about six vendors or packages.[9]

Then you subject these six to a more stringent examination, by, for example, visiting their websites and reading fact sheets about the solutions. Create a table beforehand that contains several relevant topics you want to gather more information about, such as "Does the scheduling functionality have the option to split and merge orders?" or "Does the package have extensive functionality for access security?" This helps you reduce the long list to a short list of at most three candidates. *Why not more?* Realize that every subsequent step you take is going to cost you a lot of time. The more packages you have on your short list, the more presentations you must attend, and the more responses you have to compare and assess. A single extra package on the short list can easily cost the selection team an extra two days of work.

4.4.2 The Request for Information and the Questionnaire

Send a request for information (RFI) letter to the three vendors on your short list, along with the URS document (including the requirements list with the assessment system to be filled in by the vendors) and a list of questions. The questionnaire is intended to obtain information that falls outside the scope of the MES requirements, such as the vendor's financial condition, licensing structure, strategic plans concerning the MES product, available product training, installed base, and so on. MESA provides a white paper with an excellent and detailed example of such a questionnaire.[10]

9 If you end up with too many, add one or two more detailed knockout criteria.
10 MESA, *White Paper #04: MES Software-Evaluation/Selection* (Chandler, AZ: MESA, 1996).

By the way, make sure to give the vendors enough time to respond. They're just people, after all, and they really aren't sitting by the phone all day, waiting for a potential customer to call (and if they are, then you probably haven't chosen the best of the bunch). They have other clients, they're working on other RFIs, and they have to free up people to answer the questions and prepare the presentations. Give them six weeks or so; that's a reasonable amount of time.

4.4.3 Q&A Session for the Software Vendors

To avoid having vendors keep calling with all kinds of questions, it's useful to organize a Questions and Answers session. Do it three weeks after they've received the RFI, for example, so that they've had enough time to read the documents and prepare questions, and so that they still have enough time to incorporate the new information into their presentations. This way, you minimize the amount of time you have to spend on answering questions, and you give all the vendors the same information.

4.4.4 The Vendor Presentations

And then it's time to lean back and watch the vendors sweat. If only that were true! Unfortunately, it's not yet that simple to compare different MES products, so it's still hard work. If all's gone well, the internal selection team has worked through the information the vendors have sent in detail before the vendor presentations, so you already have a first impression of what to expect. And you know which things are unclear, and what you need to ask more questions about.

Keep asking, and don't be satisfied with a Power Point presentation or a vague answer. The vendor should give a live demo of her solution; this usually requires a technical specialist. To provide a good idea of both the vendor and the package, a first presentation should easily fill at least half a workday.

4.4.5 The Reference Visit and the Proof of Concept

If the vendor presentations haven't given you enough information to make a final selection yet, you can ask for a proof of concept (PoC) or a reference visit or both. Be aware that each of these is going to cost the vendor a lot of time and effort. For a reference visit, it's also important to realize that the vendor doesn't want to bother his satisfied clients time and time again.

If you want a proof of concept, then you'll need to develop a case yourself. Describe a typical aspect of a process, or conversely, an unusual one you're not sure the vendor can support. Of course, you can't expect the vendor to work out the entire system for you, so the case must remain small and easy to digest.

It's a good idea not to organize the reference visit and the proof of concept too early in the project. After assessing the RFI documents and seeing the presentations, you'll have a much better idea what you should pay attention to during the proof of concept and the reference visit than you would in an earlier stage.

4.4.6 Selecting a Package

Hopefully, the selection team now has a good picture of the vendor and the package, and you can assess whether the licensing costs and the functionality are in proportion to one another, and whether the solution's standard functionality sufficiently meets the user requirements. Then it's time to make a choice.

4.4.7 Selecting the System Integrator

The MES software vendor can provide detailed information on the package and on the licensing structure.[11] But in most cases, the

11 Unfortunately, the licensing structure for many software vendors is very murky.

software vendor isn't the one who's actually going to implement the package,[12] which includes performing the work required for configuration and custom programming. Most MES software vendors work with local system integrators, and it's these local system integrators who have to ensure the price and the quality of the implementation.

The MES software vendor can provide an overview of system integrators who have the most experience and can provide local support. If all's gone well, the selection team has included this criterion in choosing a package. Don't forget that the more system integrators there are who support the package, the less risk you run of vendor lock-in.

There are probably very few system integrators who have sufficient knowledge of, and experience with, the package you've selected and who can offer support in the right region(s). There may be, at most, three. And those three will differ in their rates, their knowledge of and experience with implementing that specific package, their experience creating interfaces with Level 4 systems, the availability of employees with the required knowledge, their extra services such as training and implementation guidance, the presence or absence of internal competence centers, speaking the local language, twenty-four-seven support, and so on. Some of them have developed their own (domain-specific) software to supplement the package's shortcomings.

Send the request for quote (RFQ) letter to those system integrators, along with the URS. The system integrators also deserve sufficient time to write up their proposals. Again, a period of six weeks is realistic. Here, too, it's useful to offer them a Q&A session about three weeks after you've sent the RFQ.

12 Exceptions notwithstanding.

The selection team should again make sure they've read through the quotes carefully and prepared their questions before the presentations. Make sure you aren't comparing apples and oranges. There may be large differences in the project price, but what's causing them? Does one integrator also provide project management, training, and implementation guidance? Does another leave the creation of a manual up to you? To help you make a good comparison, it's useful to provide a template for the RFQ. This way, all the system integrators will provide their information in the same format.

4.5 A Happy Ending

By following these steps, you can avoid scenarios like my bathroom story. Do you want to know how that ended?

I went to another supplier, who not only sold bathrooms, but also installed them. The saleswoman did not use colored pencils, nor did she make a fancy drawing. She informed me that if I chose that specific sink, I'd need an installation kit that cost fifteen dollars. And to install the tub, I'd need a set of supporting legs that cost thirty dollars. She took two hours to help me specify the complete list.

Then we talked about possible savings. For the most expensive items, I decided to choose an alternative solution. The saleswoman advised me to choose my accessories after installation. In her experience, clients often change their minds after the bathroom's finished. So we decided to make a rough estimate for the accessories, and leave them out of the fixed-price part.

A few months later, I got the key to my new house, and they came to install the bathroom. Two men removed all the existing appliances and tiles. Then the electrician came. He brought the basic sketch that the saleswoman had made.

"You told our colleague that you want outlets near your sink. Where exactly should they go?" I pointed out the location. Based on his experience, he advised me to put them elsewhere. I agreed.

The next day, two men came to do the tiles. They also called me over when they were unsure about details on the sketch. After two weeks, all this work resulted in a beautiful bathroom, professionally installed, delivered on time and under budget.

To me, this is the perfect example of good communication between a vendor and a client. The client should make a requirements list that provides sufficiently detailed information, and the vendor should be open and clear about costs and work that aren't obvious. Don't expect the requirements list to speak for itself. Keep talking with each other in every phase. Stay in close contact, and use the URS document as the basis for communication.

You've selected the MES, and you've found a system integrator who's going to implement it. What can you expect during implementation? And what can you do in advance to prepare the internal organization for the changes to come? Read in the following chapter about the phases that an MES project comprises and what you—as the end user—definitely shouldn't forget, in order to make the adoption process as smooth as possible.

CHAPTER **5**

What's It Like, Implementing, Adopting, and Maintaining an MES?

Choosing the MES package and the system integrator was already a lot of effort. But the real work's only just beginning. The vendor and system integrator are going to build, implement, and deliver the package, and you've got work to do, too. It's a good idea to be somewhat prepared for what's coming. How long does a project like this take? What exactly is that project team doing, week after week, without giving you a glimpse of even a single screen? In the meantime, how do you ensure that internal employees are really going to use it when it's ready, and use it the way they're supposed to? And finally, who are you going to assign responsibility for maintenance once it's ready, IT or Engineering? In short: how do you turn your MES into a success?

5.1 From URS to Realization

You've written the URS, you've selected the package, and you've handed the assignment to a system integrator. Now it's time to start implementing. What's in store?

An MES project comprises a number of phases, and each phase has its own deliverables. In the MES world, it's common practice to use the **V-model** from GAMP (Good Automated Manufacturing Practice) to describe the phases and the associated documents. For many years, GAMP has been an accepted methodology for realizing validated control systems. GAMP is an initiative of the

ISPE (International Society for Pharmaceutical Engineering). The ISPE developed GAMP in order to have one common, standard interpretation of the highly abstract GxP regulations.

GAMP also has disadvantages. It's a rather strict method, with many extensive documentation and qualification steps. A new phase may only be started after the previous phase has been completed and QA has given its approval. Both forward and backward traceability are required. When, during the engineering phase, you discover that the requirements in the URS aren't complete or are wrong, the whole trajectory (URS, functional specification, design specification) has to be redone. For companies that don't have to comply with GxP regulations, all this specifying and testing seems a bit overdone.

Despite these strict rules, the GAMP V-model and its terminology are useful for defining MES project phases. If an industrial company doesn't need to comply with FDA regulations, it can always use the "GAMP light" version, meaning: only use GAMP phase names and terminology, without following GAMP's strict rules for validation and qualification.

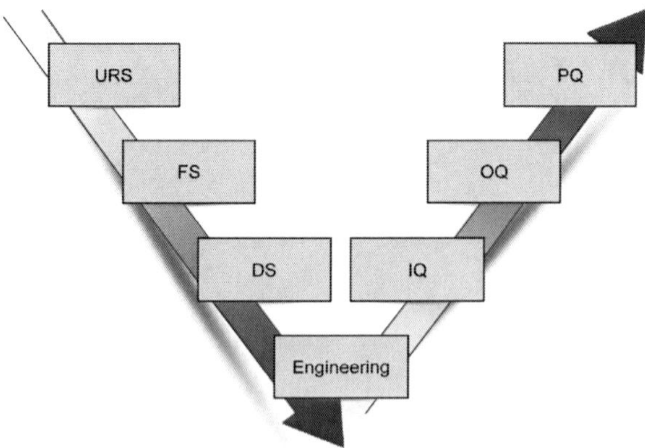

Figure 5.1 The GAMP V-model

GAMP's famous V-model (figure 5.1) makes clear that automation projects contain different phases. During the first phase, the client specifies the user requirements. After the URS phase, the (internal or external) vendor can start writing the **functional specification** (FS). In the FS document, he describes what the final system will be able to do—what he's going to deliver.

The FS is a response to the URS. The URS is described independently of the future solution. The FS describes the solution. The FS is linked to a requirements traceability matrix document to assess whether all the user requirements are met. If the supplier isn't able to provide a solution for a specific user requirement, then she has to explicitly state this. The FS must be written in such a way that the client is able to understand it.

On one hand, the FS is a means of communicating with the client. On the other hand, it's the input for the next phase, in which the vendor's going to write the **design specification** (DS, for both hardware design and software design). The DS contains a detailed, technical description of the solution. This doesn't need to be in language that the client can understand, because it isn't meant to be read by the client. The DS is read by the people who are going to build and install the solution (i.e., the vendor's engineers and programmers).

After the solution's been built, many qualification phases follow. The **installation qualification** (IQ) ensures that the hardware complies with the hardware design specification, and that it's been installed. It also ensures that all software components cooperate with each other correctly. The IQ is sometimes called a **factory acceptance test** (FAT).

The **operational qualification** (OQ) is the output of the functional specification. It's sometimes called a site acceptance test (SAT). During this phase, the target hardware at the customer's site is validated, and it's proven that the system works according to the FS.

Finally, the **performance qualification** (PQ) ensures that all the requirements in the URS were successfully realized in the production system.

5.1.1 From the Trenches: Joint ERP and MES Rollouts at Arla Foods

Based on prior experience, Arla Foods decided to combine its ERP and MES projects into one joint project. Arne Svendsen explains.[1]

> Previously, we had separate SAP and MES rollouts, which led to extra work. (The SAP screens that we had built were ultimately replaced by an automated feed from the MES.) Furthermore, it led to confusion for our on-site customers.
>
> So we decided to coordinate our ERP and MES projects. Now, we decide on the interfaces up front. We take the integration phase into account when defining the business case. We now closely link the project plans for SAP and MES. The SAP blueprint and the MES specification are developed in parallel. Testing is carefully planned, taking into account SAP deadlines. And we have a one-month "burn-in" period on site before SAP users go live. During this period, training takes place, and we have time for troubleshooting in relation to master data errors.

5.2 Adoption

My very first automation project was a SAP implementation at a large power company. I was part of a group of at least four external consultants and two internal employees, and we developed the user

1 Arne Svendsen, "Bringing Business and Manufacturing IT Together" (presentation, WBF European Conference 2008, Barcelona, November 2008).

manuals and the user training materials. We also wrote internal newsletters about the changes that were coming, and we organized information sessions and role-playing workshops to prepare the employees for adopting the system.

Later, after I became an MES consultant, I ended up at a company that had been doing PLC-SCADA systems since the dawn of time, and was now taking its first steps into the world of MES. The fact the first MES project wasn't an overwhelming success wasn't because the team lacked in technical capacity; the human factors threw a monkey wrench into the works.

Most MES system integrators originally come from the PLC-SCADA world. An important difference between SCADA projects and MES projects is that with MES, you have to deal with so many more stakeholders (just as for ERP implementations). Often, the project concerns data collection and reporting functionality, for example. The information from the system is intended for many different users, such as operators, supervisors, quality assurance employees, maintenance staff, engineers, and plant managers. The system can only be a success if these people also actually use it—and use it the right way. So it's crucial that you "knead in" the change.

"Kneading in" is a whole art in itself. It's called **implementation guidance**, and it's provided by people who usually have little understanding of technology, but all the more of communication, psychology, and other human factors. Remco Moria, a consultant at Ordina who specializes in implementation guidance, explains.[2]

> It's important that the future users, and other people in the organization who will notice the consequences, feel involved. That's why we involve them in the various implementation phases. We prepare a solid, integral change management plan

2 Remco Moria (Implementation Guidance Consultant, Ordina), interview with the author, 2007.

in order to communicate with the employees and to prepare them for the coming changes. Consider, for example, setting up a sounding board group and a users group, organizing workshops, creating training materials, and setting up a communication plan. We work with key users, because there are often too many future users to be able to reach them all at the same time. This way, the right approach to each target group also becomes clear.

From the start, the implementation guidance team looks at the whole picture. They go into the organization to interview different types of users, and ask: What issues play a role here? How are you going to maintain the system? What do you expect from it in the future? How much uptime must the system maintain? By asking different users these kinds of questions, you form a picture of the current pattern of expectations, and of the conflicting expectations within the organization. Once you know what the errors and obscurities in the expectations are, you can start to alter them. That occurs continuously, throughout the course of the project.

During implementation, it's important to pay a lot of attention to the mental changes that need to take place (by teaching, creating involvement, and generating support and acceptance, through training, workshops, and newsletters, for example). Moreover, a structural change is necessary; that is, the change has to be incorporated into the organization. The goal is to lock in the result, for example by implementing processes, setting up procedures, delegating organizational tasks and other responsibilities, functional management, and so forth.

In short: in MES projects, the system's success depends to a significant degree on "kneading in" the changes. Because they lack knowledge and experience in this area, but also so they can offer competitive rates, most system integrators restrict the communication plan to a few hours of training. It's up to the end user to fill in the rest.

5.2.1 From the Trenches: Implementation Preparation at Aviko Rixona

Sjoerd van Staveren, MES Coordinator at Aviko Rixona in the Netherlands, describes how Aviko prepared its personnel for the new system.[3]

> At the time, I paid a great deal of attention to preparing the operators for the changes they were going to encounter. I talked extensively with them, and I gave several presentations in meetings and departmental gatherings. It didn't truly become real for these men and women, however, until they got behind the wheel themselves.
>
> We'd built a test environment, including an operator station. We simulated the process for a filling and packing line from starting an order, through displaying OEE data, to making production responses and closing out the order. At the time, in 2002, the fact that several of these people had never used a PC also played a role. That was something we had to take into account, and I think you still do today. It happens less and less frequently, but now and then you still run across people who don't do much with computers at home or at work, and you have to help them get over being gun-shy.

5.3 Maintenance

A few years ago, I met a production manager who told me she wanted to purchase an MES solution. She had contacted her

3 Sjoerd van Staveren (MES Coordinator, Aviko Rixona), interview with the author, December 2008.

engineers for help, but they sent her to the IT department. When she went to the IT department, she got sent back to Engineering. Her story made me curious. I didn't have a clear answer myself to the question which of the two should have helped her. Of course, Production is the *owner* of the system, but where can they go for project assistance, system support, and system maintenance?

Traditionally, IT departments handle the ERP and other systems used in the office. The functionality in these systems belongs largely on ISA-95 Level 4. These IT people "speak" Java, .Net, and other programming languages. Engineers[4] are working in a completely different world, namely, the world of instrumentation, PLC, SCADA, and DCS, with programming methods like ladder logic, SFC, and function blocks. They're active on ISA-95 Level 2 and lower. But who takes care of automating activities on ISA-95 Level 3; in other words, MES?

5.3.1 Engineering or IT?

Over time, traditional Level 2 process control system vendors have offered more and more Level 3 functionality, such as historians, standard production reports, and recipe management modules. These systems are closely integrated with the systems on Level 2, so it's essential to have a thorough knowledge of the process and its automation. This suggests that the engineers should be the ones to support MES and execute MES projects.

On the other hand, there are reasons to think that IT should support MES. The MES software vendors have based their historians, reporting functionality, and recipe management on the operating systems, programming languages, and network protocols that are characteristic of IT environments.

4 I call this *Engineering*, but it might be called *Automation* in your company, or it may be part of Maintenance.

When I asked two of my clients who in their companies was responsible for supporting MES projects and maintaining MES systems, I got contradictory responses. I decided to do some research, in a very informal, non-scientific way. The purpose was to get a first impression of the current situation at different clients, to collect a variety of opinions, and to list some advantages and disadvantages of MES support by IT and MES support by Engineering.

"It can't be a coincidence that we keep addressing the same topics," responded one MES project leader after I sent him a short list of questions. "I've just become a member of a European working group to investigate this very same subject. It appears that each of our plants in Europe has a different approach. The purpose of the working group is to develop a recommendation on where to place the boundary between the different disciplines and the related organization. The final goal is to develop a common approach."

Other people also responded enthusiastically when I asked them to give their opinions. It appeared to be a hot topic that needed clarification.

I sent my questions to eighteen end users from my personal network.[5] Fifteen of them responded before the deadline. (Please note that this was a small and very informal survey. There were too few participants to consider this as representative.) The respondents all work for international companies with headquarters in Europe and in the United States. Most of the respondents work in the pharmaceutical and chemical industries, but some individuals work in food, tobacco, energy, and biotechnology. Almost a quarter of them are active within Level 4 IT; almost half are active within Level 3 IT; almost a quarter are active within Engineering, and one works for Production. Note

5 The following people participated in this survey: Wouter Amersfoort, Paul Bekkers, Alex van Delft, Charlie Gifford, Charlotta Johnson, Henk Leegwater, Valentijn de Leeuw, Remco Moria, Elena Pileri, Paul de Ruigh, Erwin Winkel, Martin Zeller, and others who chose to respond anonymously.

that seven of them say they're active within more than one of these disciplines. Only 62 percent of these companies differentiate between "Level 4 IT" and "Level 3 IT" (see figure 5.2).

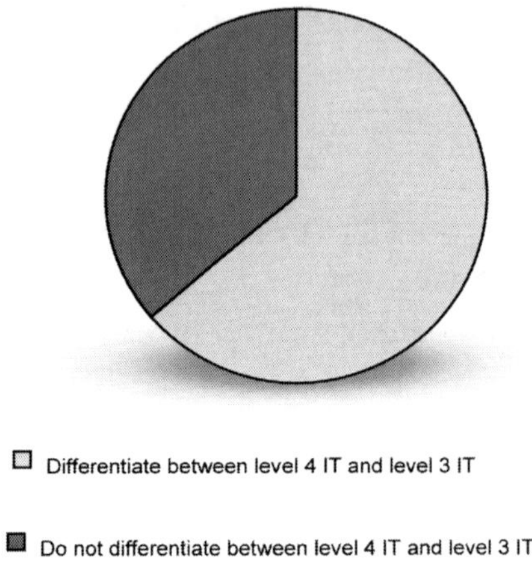

Figure 5.2 Does your company differentiate between Level 4 IT and Level 3 IT?

One of the respondents told me his company decided, around the year 2000, to develop its own custom MES solution, because at the time no mature MES solutions were available on the market. They established a sub-department within IT that was responsible for MES. It was a necessary step, because they needed to build interfaces with the process control systems.

It turned out not to be so easy to build a good relationship between the IT sub-department and the Engineering department. Neither the IT people nor the engineers trusted the other enough. The result was a struggle for power. Only years later did the situation improve.

In another company, IT started fiddling with MES about twenty years ago, as a kind of curiosity. This mostly concerned data collection. They didn't align with Engineering, which had a very negative impact on process automation. The MES system had to be removed in order to get the production process started again.

A vague division of responsibilities has been an issue, or is still an issue, within many companies. Their past MES efforts have resulted in a hodgepodge of implementation methods, system functionality, system boundaries, responsibilities, and so on.

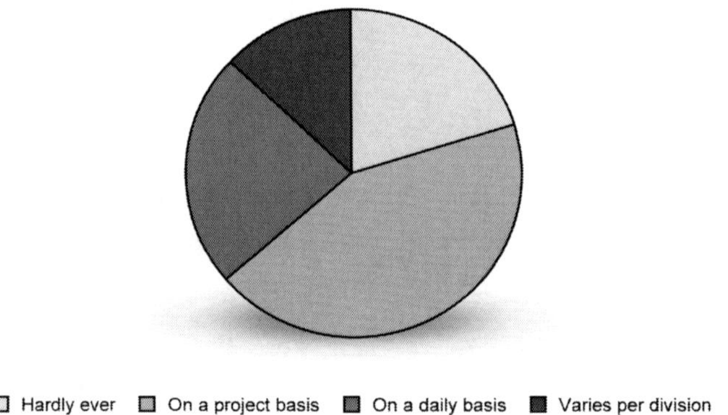

Figure 5.3 How closely are IT and Engineering working together in your company?

When I asked respondents how closely IT and Engineering are working together in their companies, 20 percent told me there's no collaboration at all (see figure 5.3). Forty-three percent said they work together on projects, and 23 percent said they work together closely, on a daily basis. Thirteen percent couldn't give a single answer to this question, because there are too many differences between divisions.

"In my opinion, IT and Engineering don't collaborate in our company," one engineer said. "Even worse, they're antagonistic to one another. Although the technical gap between IT and Engineering is getting smaller, the chasm between the departments is still huge. I think this heavily impacts the success of our MES implementations."

"Generally speaking, we consider 'IT functionality' something that's supported by the IT staff, whereas the Engineering department supports DCS, SCADA, and PLCs. Because MES systems contain both, both departments will have to be involved in MES projects," said one respondent. "In practice, this means that 80 percent of the work is carried out by IT, and 20 percent by Engineering."

One respondent remarked that, within MES projects, their IT and Engineering departments work together closely, but collaboration needs to improve for MES systems support.

When there's daily collaboration, IT and Engineering usually report to the same entity. Several respondents reported that the relationship between IT and Engineering greatly improved from the moment they were forced to work on a single, integrated architecture. This led to greater clarity on roles and responsibilities, and they started to trust each other more. Establishing a Level 3 IT department ("Manufacturing IT") also seems to positively impact collaboration. You need a bridge, and that bridge can be built by establishing a dedicated Manufacturing IT group that combines IT skills with automation skills.

5.3.2 What Can IT and Engineering Learn from Each Other?

The respondents mentioned several strong points and weak points for both IT and Engineering, when asked which party would better support MES projects and maintain MES systems.

5.3.2.1 IT's Strong Points

Of course, one of IT's strong points is their knowledge of information technologies such as Ethernet, the IT infrastructure, networks, protocols, topologies, databases, data management, security and software development technologies, tools, and methods. An MES is an application that's oriented toward interfaces and databases. Databases are not part of a technical engineering environment.

"IT departments are larger and can afford to have more specialized knowledge," said one respondent. "In most companies, Engineering has been stripped to the bone and fragmented to provide support for manufacturing systems and equipment at diverse sites. This severely limits the talent pool at any single site."

IT is also reported to be more professional when it comes to system maintenance. They're better at systems monitoring, making backups, and developing system documentation. Engineering, on the contrary, provides support "on the side." It isn't their core task, and they handle it less professionally. Furthermore, IT is used to working with several disciplines. Like ERP systems, MES systems are used in different fields, so this experience is helpful.

A very frequently mentioned advantage of using IT for MES support is its central location within the company. From its vantage point, IT has an overview of the whole site, or even the entire company. So they're familiar with the infrastructure MES uses, and they're already responsible for the desktops MES uses. Moreover, they know the other business processes and systems with which MES has to exchange information. From this central position, they have a better basis for harmonization and standardization, and for avoiding redundant functionality. As a result, they're capable of developing a corporate MES strategy and integrating it into the corporate IT strategy. This integration leads to synergy among plants; implementation knowledge could be reused from one plant to the next. Ultimately, it results in cost efficiency.

Engineers don't have this bird's-eye view. Historically, they focus on local projects, whose scope doesn't reach beyond a production machine or a production line. One of the respondents reported that his company's Engineering department had once implemented OEE functionality for a production line, which—from a business perspective—wasn't interesting at all, because that line wasn't a bottleneck. So, unnecessary investments can be the result when they're not first assessed at a higher level.

The idea that Engineering doesn't have enough information to be able to assess the MES business case is supported by another interviewee. "How can they know what the costs and benefits of integrating MES with LIMS are? And how can they know whether material tracking and tracing would be best supported by the warehouse management system, the ERP system, or the MES?"

In short, there are many advantages when IT is responsible for supporting MES projects and maintaining MES systems. This suggests that IT should hold the reins. But let's see if there are any significant advantages when Engineering carries out these tasks.

5.3.2.2 Engineering's Strong Points

Knowledge of, and affinity with, production is a very important characteristic of Engineering. Even if information technologies continue to exert ever more influence on process automation, you'll always need detailed knowledge about the production process, the product specifications, and the limitations of the process and the equipment. Typically, the heart of the MES system is a model of the production process, and that model must be aligned with the Level 2 systems.

Especially when the production process comprises a mixture of manual steps (supported by MES) and automated steps (supported by the Level 2 system), the MES will have to be in close harmony with Level 2 systems, and this is something Engineering can take care of.

IT is not renowned for its extensive knowledge of the production process. Both physically and in terms of experience, they're removed from the processes with which MES must be integrated.

"With an MES system, it's essential to fully understand the processes about which MES is reporting. If you don't, you're reporting rubbish!" said one interviewee.

Another said, "It's not feasible to train IT people to the extent that they'll be able to deal with this. The biggest challenge for an IT person is to think like a production person. For example, reliability within the four walls of a production plant is much more important than corporate efficiency or being able to work remotely. You can't decide just like that on a Friday night to shut down the system for a few hours of maintenance. Engineers are reputed to understand this much better than IT people. They're much closer to the reality of the production process. Engineers usually report directly to production managers, and they're part of what companies call 'Production.'"

"Engineers are familiar with production requirements and are willing to develop solutions meeting those requirements. Engineering doesn't just come into play after a functional specification is ready, but helps create a user requirements specification and develops the functional specification on its own," explained a respondent who works for an Engineering department. Engineering can also provide twenty-four-seven support, whereas that's not the way IT tends to work.

Engineering projects have a direct impact on production processes. Often, these projects can only be realized by stopping production. These production stops must be as short as possible. This is an aspect with which IT people usually aren't familiar.

Moreover, corporate IT and security policies can't be applied to production systems the same way they're applied to enterprise systems (see section 5.4).

5.3.3 So Who Should Be in Charge?

In short: both Engineering and IT possess important knowledge and skills for MES projects and MES systems support. When it comes to assigning responsibility for these tasks, many companies seem to be in a situation that's grown organically instead of resulting from a conscious, strategic decision. The ones that did make a conscious decision, didn't all make the same decision. There are still many diverse options, and nothing indicates that one scenario is better than the other.

For example, you could put IT in charge and have them communicate with Engineering on a project-by-project basis. You could do the same thing the other way around. Another option is to establish a dedicated Manufacturing IT department. Yet another possibility is to put IT and Engineering under one roof, by merging the departments.

The questions I raised in my survey are burning, and we'd all like to find the answer to them. But there are a lot of different opinions, and even now, hardly any information about best practices is available. So beware! If someone tells you "This is the best way to do it," you're probably listening to a *personal opinion* and not a *scientifically proven fact.*

Until more information about best practices is available, I can only make you aware of this point of concern, and advise you to pay special attention to where you place responsibility for system support when you introduce MES in your company.

5.3.3.1 From the Trenches: Bridging the Gap between Engineering and IT at Janssen Pharmaceutica

The company Janssen Pharmaceutica also found itself confronted with the question, "Is MES an IT system, or does it belong within the Engineering scope?"[6]

> Up to that point, IT and Engineering had very much been separate organizations, where IT typically was responsible for

6 Kurt Werckx and Carl Van Laer, "Approach to a single MES strategy within a global environment" (presentation, WBF European Conference 2008, Barcelona, November 2008).

all systems other than the DCS systems and the manufacturing reporting applications. Looking at the scope of our MES strategy project, we immediately recognized the possible negative impact this separation might have on the project. So this was an issue that we had to deal with. We resolved it by forming project teams with both IT and Engineering representation in all layers of the project.

<table>
<tr><th colspan="5">Project Sponsors</th></tr>
<tr><td colspan="5">Manufacturing VP – IT VP</td></tr>
<tr><td colspan="2">Steering committee</td><td colspan="3">Global Core Team</td></tr>
<tr><td colspan="2">• Senior Director Manufacturing
• Senior Director Quality & Compliance
• Senior Engineering Director
• IT Director</td><td colspan="3">• Manufacturing representative
• QA representative
• Engineering representative and project manager
• IM [IT] representative and project manager
• MES consultant</td></tr>
<tr><td colspan="5">Local Site Representatives</td></tr>
<tr><td>Area/Site</td><td>Noramco</td><td>Cork</td><td>Geel</td><td>Schaffhausen</td></tr>
<tr><td>Engineering</td><td>Xxxx</td><td>Yyyy</td><td>Zzzz</td><td>Aaaa</td></tr>
<tr><td>IM [IT]</td><td>Xxx2</td><td>Yyy2</td><td>Zzz2</td><td>Aaa2</td></tr>
<tr><td>Manufacturing</td><td>Xxx3</td><td>Yyy3</td><td>Zzz3</td><td>Aaa3</td></tr>
<tr><td>Quality</td><td>Xxx4</td><td>Yyy4</td><td>Zzz4</td><td>Aaa4</td></tr>
</table>

Concerning maintenance of MES after going live we were aware that, although we use a similar MES template on all sites worldwide, still interaction is required to local, plant-specific DCS systems. So there is a need for a local organization being able to deliver support. For us, integration of the MES with the local DCS system is essential to create a full electronic batch record. If this integration is required, there is more often a need to have the MES system supported by IT as well as Engineering because Engineering owns the DCS and DCS upgrade programs.

5.4 Security

Last but not least, this chapter isn't complete without a comment on MES security. MES is located in what ISA-95 calls the Control Domain. The points of concern in control systems security differ from those in enterprise systems security. You'll hear the acronym **CIA**, which in this case stands for Confidentiality, Integrity, and Availability. In enterprise automation, information confidentiality is the most important point of concern (see figure 5.4). Then comes integrity, and only then does information availability become important.

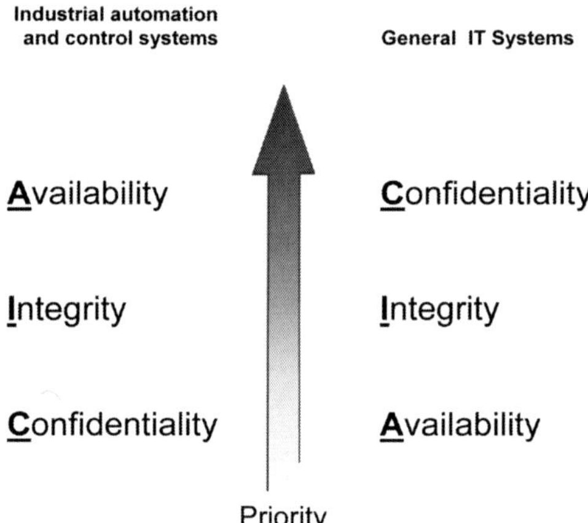

Figure 5.4 The ISA99 standard makes clear that the priorities for control systems security are different from those for ERP systems security

In production automation, it's just the other way around. Security for these systems has as its primary goal to guarantee availability of all parts of the system. Certain hazards go hand in hand with the machines that are controlled, monitored, or otherwise affected by the automated systems. For that reason, integrity takes a back seat. And confidentiality is usually even less important, because these are raw data that only become confidential within a certain context.

The required response time is also an important difference. Control system response times must sometimes be accurate down to the millisecond, while enterprise systems work just fine with a response time of one or more seconds.

In short: company-wide IT procedures shouldn't be naively applied to the MES. ISA has developed a standard for control system security: ISA-99. This is a useful tool; it adds expertise from the control system domain to already widely available IT security technologies. The standard provides workable, practical solutions for control systems, without causing more damage than the things you're trying to prevent will cause!

Now you have an idea of what an MES is, what it can do for you, and how to select one, build it, implement it, and put it to use. If there's a business case for MES in your company, you can get started. Good luck!

For companies with multiple plants who first want to know whether and how you can use one and the same MES at different sites, the following chapter contains several tips and real-world examples.

CHAPTER **6**

Can We Roll Out the Same MES in All Our Plants?

Many companies have one central ERP system in use that supports the processes at different sites. For MES, it's more logical to install the system locally. But that doesn't change the fact that it can be advantageous to use "copies" of the same MES at the various plants, instead of developing a separate system for each plant. What are the benefits? How similar do the plants have to be in order to use the same functionality? How do you develop an "MES template" step by step? And once you've put it into service, how do you make sure the versions at the different sites stay aligned? In short: how do you develop, implement, and maintain a multi-site MES template?

6.1 Pros and Cons of a Multi-site MES Template

Many companies with multiple plants are confronted with a profusion of information systems, including all the associated drawbacks. For example, you have to maintain several different interfaces. You're also dependent on employees with knowledge of diverse—sometimes exotic—technologies. And economies of scale and reuse are out of the question.

That all changes when you decide to use one and the same MES at the individual sites. You invest just once in selecting, specifying, and building the system, and then you roll it out at possibly dozens of plants. A central competence center houses the—in this case limited amount of—required knowledge and experience.

Another advantage is that the information spread across various plants is now standardized, so that it forms a better basis for plant benchmarking, and for exchanging knowledge about the processes.

Last but not least, you're in a much stronger position to negotiate the price.

Unfortunately, none of this alters the fact that using the same MES in all your plants also has disadvantages. Suppose the MES vendor goes bankrupt, or stops developing and supporting the product; every plant will feel the blow. So be sure you come to an understanding with the vendor on source code availability for this kind of emergency. If you have the source code, you can adapt and expand the MES yourself as needed.

For that matter, all sites will be plagued by the same problems in less serious situations, too, such as bugs and poorly implemented functionality. That said, the advantages usually outweigh the disadvantages.

6.2 Multi-site MES Template Attainability

When your company first started using its ERP system, the process probably went hand-in-hand with changes to procedures and working methods. The information system didn't adapt to your customary way of working; your organization adapted to the information system. It isn't so black-and-white with MES.

Of course, you can adapt plant procedures and routines to a certain degree, but the production lines are the way they are. You don't just convert a packaging line because the MES can't properly support it otherwise. To what degree is using one and the same MES in different plants attainable, then?

If one plant makes snacks and the other soft drinks, it may not be attainable. But if one group of plants manufactures roofing tiles from clay and another group manufactures them from concrete, it undoubtedly is. You can probably distinguish a number of plant groups that have the same qualities in terms of production process and product families. For example, do they execute batch, discrete, or continuous production processes? Do they have similar complexity in terms of the number of runs per day or per week, the number of different items they produce, or the degree to which flexibility is required? By asking such questions, you can define a number of logical groupings and determine per group, in a more thorough analysis, what the differences and similarities are.

Note also that the boundary between ERP and MES at all the sites involved must be drawn the same way in order to employ an MES template (see figure 6.2). Then the interfaces between ERP and MES can be made to comply with the ISA-95 series of standards. That way, you avoid having to invest time and money in various custom interfaces. And you don't have to develop custom functionality at each site to fill up the holes between MES and ERP that are caused by the—in that case—erratic ERP border (see figure 6.1).

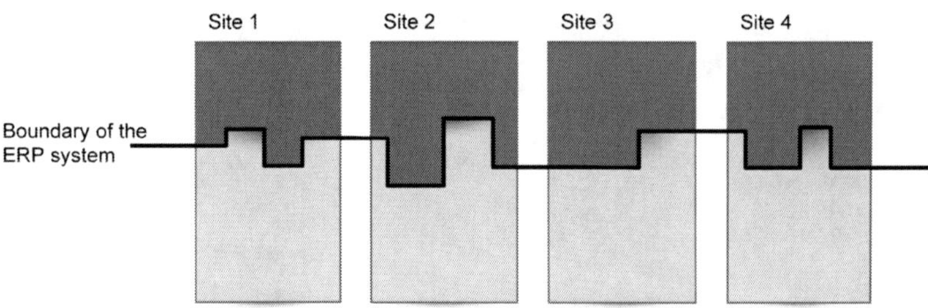

Figure 6.1 Boundary of the ERP system (not standardized across sites): this situation makes integrating a multi-site MES template with the ERP system more difficult

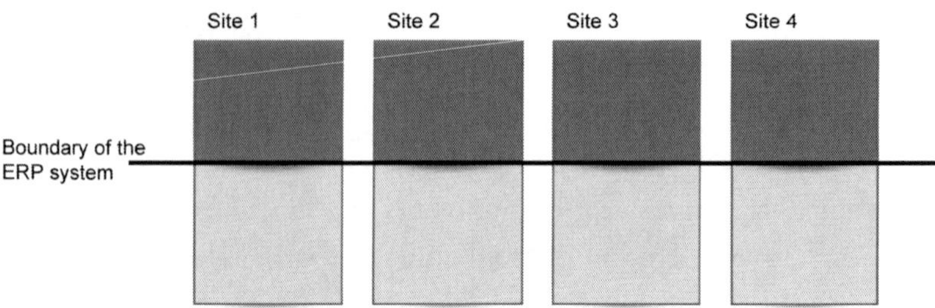

Figure 6.2 Boundary of the ERP system (compliant with the ISA-95 series): this forms a better basis for integrating a multi-site MES template with the ERP system

6.3 Road Map for Template Development and Implementation

The first time I had to write the requirements for a multi-site MES template, I visited all the plants one by one to talk with the users and merge all their requirements into a single URS for MES. That's doable when you're talking about three or four plants, but what if the division has over thirty plants?

6.3.1 Developing the Template

In the case of those more than thirty plants, we decided to form a working group, in which each group of plants (three groups in total) was represented by one plant manager. The working group also comprised an internal project leader, the IT manager, a representative from the future internal MES competence center, an external MES consultant who led the meetings and workshops, an external MES consultant who took notes and wrote the documents, and an external project leader. For a few sessions, we called in the help of an internal ERP interface specialist. The working group reported to the steering committee.

We had a tight schedule, because we wanted to have a price estimate before the budget reviews. To be able to make a detailed quote for an MES implementation, a vendor needs information on not only the desired functionality, but also such aspects as how many licenses are needed, which PLCs the MES needs to interface with, how many production lines are within the scope, and so on. But these aren't topics that belong in a multi-site MES template. After all, they're site-specific. So we decided to divide the project into two phases: the definition of the multi-site MES template, and the drafting of the site-specific URS for a pilot plant. The multi-site MES template definition would describe all the standard functionality, while the site-specific URS would provide extra information so the vendor could make a quote for implementation at that site. The pilot plant quote would then serve as a guideline for the cost of implementation at the remaining sites. Table 6.1 shows the road map for defining the multi-site MES template.

Table 6.1 Road map for defining the multi-site MES template[1]

Date	Step	Action	Explanation
Week 1	1	Kickoff	Discuss agenda, business drivers, scope, time schedule, and so on. Explain the ISA-95 methodology.
Week 1	2	Workshop preparation	Prepare templates for workshops. Analyze existing documents and describe functionality of existing MES systems that need to be replaced.
Week 2	3	Workshop day 1	Cover ISA-95 physical hierarchy, process segments, equipment, personnel, material, products. List characteristic examples and list characteristic differences between plants that the solution will have to be able to deal with.
Week 2	4	Workshop day 2	Cover ISA-95 production scheduling, production resource management, production dispatching, production execution management. List characteristic examples of schedules, dispatch lists, and resource management, and list characteristic differences between plants that the solution will have to be able to deal with. Describe how the system will be used during production execution.
Week 2	5	Process results from workshop days 1 and 2	Consultant writes first chapters of the draft document and requirements list.

1 This road map was developed by consultants at TASK[24] and Ordina Belgium.

Date	Step	Action	Explanation
Week 3	6	Review results from workshop days 1 and 2	The steering committee and the working group review the results and provide feedback. The plant managers discuss the results with their employees.
Week 3	7	Workshop day 3	Cover ISA-95 production data collection, production tracking, production performance analysis. Describe typical data that need to be collected from Level 2 process control systems, operator comments, and so forth. Describe typical interfaces with Level 2 control systems. Describe typical standard reports and their contents. Describe typical analysis functionality, such as OEE. Describe feedback to Level 4 ERP (interface data). List characteristic examples and list characteristic differences between plants that the solution will have to be able to deal with.
Week 3	8	Workshop day 4	Cover ISA-95 interfaces: quality tests, maintenance, inventory. Describe requirements for interfaces with quality, maintenance, and inventory systems / departments. Cover ISA-95 Other activities. Describe requirements for security, regulatory compliance, preferred database, preferred interface technology, and so on.
Week 3	9	Process results from workshop days 3 and 4	Consultant writes next chapters of the draft document and requirements list.
Week 4	10	Review results from workshop days 3 and 4	The steering committee and the working group review the results and provide feedback. The plant managers discuss the results with their employees.
Week 4	11	Process feedback on results from workshop days 1 and 2	The consultant and the working group process the feedback from workshop days 1 and 2.
Week 4	12	Process feedback on results from workshop day 3 and 4	The consultant and the working group process the feedback from workshop days 3 and 4.
Week 5	13	Write penultimate document draft	The consultant integrates all the information in a draft multi-site MES template specification.
Week 5	14	Review penultimate document draft	The working group reviews the draft document and the plant managers discuss it with their employees.
Week 6	15	Write final document	The consultant writes the final document; the last comments are processed and the document is handed over to the steering committee.
Week 7	16	Assess final document and requirements	The steering committee assesses the final document and requirements list. The consultant processes the last comments.

It was an extremely tight schedule, which we met thanks to the team members' enthusiasm. They understood the project's importance, and rescheduled other appointments to make time for it. That's very important in this type of project. People from different countries with top positions and overcrowded schedules have to get this job done together. If they don't make it a priority, it will fizzle.

6.3.2 The Per-site URS

The rest of an MES project proceeds more or less according to the steps in chapter 4. After you've selected the pilot plant, the site-specific user requirements can be determined there.

Because you already know what functionality you need, you can select an MES product at the same time. Figure 6.3 makes clear which steps can be carried out in parallel.[2]

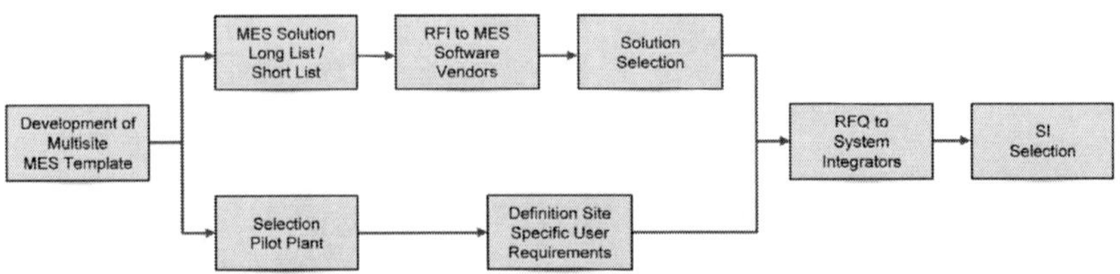

Figure 6.3 After developing the multi-site MES template, several steps can be carried out in parallel

6.4 Template Maintenance

Once the MES system is in use at multiple sites, it's quite a feat to manage it. The implementations can't deviate much from one another, or the advantages of the multi-site template will be lost.

2 This schedule assumes that you first select the product, then the system integrator. However, you could select a preferred system integrator first, and let him or her propose a product; or make combinations of preferred MES products and preferred system integrators. The best approach depends on things such as: In how many countries will you roll out the MES? How many different system integrators can provide support for the MES you've selected?

How do you keep that under control? Consultant Nik Vandenweyer at Ordina has gained a great deal of experience with this at Noble, part of the Arcelor Mittal group. In the following interview, he gives several tips for setting up an MES competence center.[3]

> Noble has fully outsourced the IT for the MES. At first, we were an internal department, but we were sold to Ordina. That way, the plants can concentrate on their primary tasks.
>
> We [The competence center] are responsible for requirements analysis, the functional specifications, the conceptual design of the IT architecture, processing change requests, implementing changes, performing maintenance, and so forth.
>
> Noble chose a custom solution with the intention of implementing it at multiple sites. To keep it as generic as possible, we based it on the ISA-95 data models.

Why the choice of a single system for multiple sites?

> Yves Detroch, the person responsible for [Arcelor Mittal's steel service] centers, wanted to use the same IT solution at all [of them]—primarily because that's cost-efficient. That way, he could spread the financial investment for development of a custom solution over multiple sites. Niko Van Der Borght, VP of Corporate Projects at Noble International Europe and Asia, continued this approach in realizing the MES solution for all Noble's laser-welding companies.
>
> There were also other advantages. Thanks to the MES template and the associated MES competence center, new plants can more quickly take part in global operations.

3 Nik Vandenweyer (MES consultant, Ordina Belgium), interview with the author, 2008.

Moreover, Noble wanted to extract information from the plants at the group level. Many plants reported in Excel. That can be manipulated; the MES template can't. By reporting through the template, you can benchmark plants based on reliable data. For example, corporate management can compare logistic data, production data, or OEE data.

What are the points of concern in setting up and utilizing a competence center?

Make very clear agreements on authority and responsibilities. You have to have a foundation of trust, because the company will become dependent on the competence center to a certain degree. The external party also helps determine the direction you'll take. Together, you have a vision of what IT should provide the company. During acquisitions and when starting up new companies, our competence center also brings part of the business knowledge to the new company. It's a very close collaboration.

What's the right moment to begin a competence center?

As soon as you define the business requirements, someone from the future competence center should be involved. Someone who knows the business well, and who has a good relationship with both the internal organization and the team that's going to realize the solution, such as the solution provider's team. At Ordina, we knew precisely what the company did. We communicated well with the company and with the IT group. We knew where the company wanted to go with the IT technology. In short: knowledge of the technology and of the business determines the makeup of the competence center.

If you aren't involved in the MES starting with the analysis of the user requirements, it's a lot harder to understand the background factors and the functional choices.

How do you staff a competence center? What competencies does it need?

The competence center is run by a competence leader. He or she has to be well informed on both functional and IT aspects. And he also has to have a clear vision. The leader is assisted by people with a strong IT background, and by senior engineers. You can expect the IT people to know everything about technological evolutions, programming languages, databases, and reporting tools. They primarily provide the technological knowledge. You also need people who are familiar with the ERP environment, in order to realize the integration. They don't necessarily have to be part of the competence center, but they do have to be available in the organization.

The senior engineers know the processes in the plants, the SCADA systems, the local historians, and the associated technologies. So the business knowledge resides primarily with the competence leader and the senior engineers, not the IT staff.

You also need people to carry out the implementation; for example, people who create the interfaces with the ERP and with the process control systems, who understand how you calculate the OEE, or who know how you configure other functions in the system.

At Noble, the MES competence leader indicates what should happen in the long term, and how problems will be solved. He also determines the budgets. He's assisted by two senior staff members: one who knows the production environment, and a second who understands the logistic flows. We also have someone in our team who's thoroughly familiar with the historical data and the line integration, and someone who's well informed on the integration with SAP. Finally, we have a pool of three junior staff who are guided by others. We currently support nine plants with this team.

How do the competence center and the individual plants communicate?

On the corporate level, the production managers are the most important contact people for the competence leader. Then we talk with the local engineers and users about their desires and needs.

One senior staff member is responsible for maintaining the applications in the various plants, and for new releases. He sets the priorities in consultation with the competence leader. Someone in the maintenance team also provides first-line support. For twenty-four-hour service, we set up a rotation system. At Noble, we aren't on standby twenty-four hours a day. Someone's always present during office hours. If there aren't any user requests, he keeps himself occupied with bug fixes or change requests. But he responds immediately to issues. On nights and weekends, someone's available for emergencies. That takes place remotely, because the plants also have their own first-line support.

Whoever's locally responsible for the plant's first-line support has to be well versed in the package's functionality. This person acts as a filter to prevent users' functional questions from ending up at the help desk. If he can't solve the problem himself, he calls the competence center. The on-site first-line support is often carried out by the shift leaders. It's important to invest in training these key users.

For a new plant, we plan a doubling of the maintenance budget for the first year. So users have a year to make novice mistakes and ask questions. Our experience is that this pretty much stabilizes after six months to a year.

How do you schedule new releases? Is it best to roll out everything at the same time? And what if not everyone needs all the functions provided?

You're best off rolling out new releases in all plants at the same time. That simplifies version control. We release a new version every three months. If you're also working on a rollout in a new plant in the meantime, you have to streamline [these projects]. It's better to avoid having different versions in circulation, because that makes the maintenance costs much higher, and you lose the benefits of the MES template. If we get a change request from one plant, we first submit it to the corporate committee. Only after acceptance do we realize the change, in all plants.

That doesn't mean that all plants also use all the available modules. They do all use the core functionality. But some make more use of the quality functionality; others, the maintenance or logistics functions. We can activate or deactivate components. That's useful: plants that have just started using the system have a greater need for root cause analysis tools, for example; others, for optimization functionality.

How do you avoid having people build things locally, so that the local functionality deviates from the template?

Arcelor Mittal doesn't allow the development of purely local MES functionality. The plants are responsible for their own PC parks and for the network infrastructure. The IT person in charge must be able to provide first-line support, but has no role in MES maintenance. The MES has to be flexible when exporting data, so that local reports can be made using MS Excel or another tool. Data maintenance is only allowed within the MES. That prevents data duplication. You also avoid having employees spend too much time on data maintenance. And, ultimately, you have access to more reliable data, also for analyses.

Competence centers often swing from peaks to valleys in terms of workload. How do you deal with that?

If it's very busy, the highest priority is that the business keeps operating. You solve problem situations like viruses immediately. In addition, employees also contribute to projects and to realizing changes.

Investment projects also have high priority. If a new plant or a new line gets added, IT mustn't be the bottleneck. The startup has a higher priority than IT projects such as migrating to a new SQL version.

You realize the rest depending on the team's capacity. To complete a project more quickly, you can add extra resources, for example, or add junior staff members. If it's calm, you can lend people to other projects. You have to lend them on flexible terms, because you want to keep the knowledge available.

How do you distribute costs for maintenance and development?

> It's usually best to assign and account for everything centrally. That way, you keep control of the central functionality, and you don't make unnecessary local investments. But if it can't be done centrally, then in any case it's the competence center's responsibility to communicate the central vision and enforce cost efficiency. Aside from that, you must in all cases be able to make decentralized reports; that is, to assign the competence center's man hours to specific plants when relevant. If we do something for several local plants at the same time, we create orders that are assigned to that group of plants.

How do you assign budgets for things like maintenance and new functionality?

> We use a yearly budget for change requests. They're evaluated by a change board. We make the budget for change requests the same as the budget for support. Support includes, among other things, bug fixes, releases to the plants, workshops, and data consistency validation. Change requests are functional changes that require fewer than a set number of days—for example, ten or twenty. If they need longer, we call it a project. Projects get their own budgets and follow an acceptance procedure.

Do a multi-site MES template and an MES competence center really save money?

> The lower IT costs are absolutely an important advantage in a multi-site rollout. That's a big difference compared with MES implementations and maintenance for a single site. It [a multi-site rollout] leads to reuse, and you standardize the requests.

Per request, you only have to develop a solution once, which you then implement in every plant. That produces enormous savings.

There are also important advantages for new sites. They can make use of accumulated best practices. You only have to look at the exceptions, and set up additional functions where necessary. You also save time on data consolidation. For example, each plant submits the safety stock levels in the same way, and the efficiency data. That makes it possible to keep an eye on the plants' delivery reliability. After the cost savings, reliable benchmarking is the most important benefit.

And of course every MES has its own business case, independent of its use at multiple sites.

Are there other important points of concern for setting up and managing a competence center?

Competence center employees generally have a difficult job. They work on international projects, they're often away from home, and they work overtime. At the same time, they often have years of experience and crucial knowledge. If you want to keep them on board, it's a good idea to implement job rotation. For example, let them work alternately on projects and in support roles. The rotation helps them more easily grow into a higher level, and junior staff members master the details more quickly.

Training junior members is important: specialists also want to work close to home now and then. That's why we give the lead position to someone different with every new project. Job rotation helps you keep minds sharp and create team spirit.

> That way, people in support roles better understand the tasks of their colleagues in projects, and they know what's going on in the various plants.
>
> We pay a lot of attention to transfer of knowledge. Together, we make a specific document for every plant. Each plant has its own work flows and profiles, for example. We create s pecific training documents that the plant leader can then translate into the local language. We facilitate, but they create the final user manuals. They're the owners.

6.4.1 From the Trenches: Advantages of a Multi-site Template at Noble

Niko Van Der Borght, VP of Corporate Projects at Noble International, told me what a multi-site template provided his company.[4]

> The IT savings from implementing a multi-site MES template are of secondary importance compared with the operational savings. Having reliable, comparable data allows you to benchmark and to implement best practices. You can base your actions on facts, which moreover are accepted by everyone; all heads are pointed in the same direction.

6.4.2 From the Trenches: MES Project Challenges and Savings Realized at Janssen Pharmaceutica

At the WBF European conference in 2008, Carl van Laer and Kurt Werckx discussed the challenges and advantages that Janssen

4 Niko Van Der Borght (VP of Corporate Projects, Noble International Europe and Asia), interview with the author, January 2009.

Pharmaceutica encountered during its first MES project.[5]

> The MES strategy definition went well, but the closer we got to implementation, the more anxiety there seemed to be, for fear of losing local flexibility. Users were wondering, "What will be forced on us? What can we still configure? How will this fit into our small organization?" We solved this by using a project organization similar to the one we used during our MES strategy definition. Involvement of all sites and all important stakeholders is required when defining your global MES core template.
>
> Another challenge was to identify the best project organization when executing a site implementation. We needed people with thorough knowledge of the global MES template, as well as people able to reflect local aspects, in order to make the implementation successful. By following this approach, we were also able to ensure that maintenance and first-line support for the MES system can be performed locally.
>
> The same approach was followed by our colleagues in the bio business. They were able to reuse a lot of the work and templates we created. This dramatically reduced the time required to define their MES strategy, and they spent only 50 percent on external consultancy compared to the first project.
>
> Moreover, if each individual site had selected the same MES vendor instead of going for corporate contracts, they would have paid almost 50 percent more than what was negotiated by corporate purchasing. The same is true for maintenance.

5 Carl van Laer and Kurt Werckx, "Approach to a Single MES Strategy within a Global Environment" (presentation, WBF European conference, Barcelona, November 2008).

6.5 From the Trenches: Multi Site MES Template Development and Maintenance at Arla Foods

The best practices outlined in the sections above stem from my own professional experience and that of consultants with whom I've worked. But you can also do things completely differently, of course, as can be seen from the following conversation I had in September 2008 with Arne Svendsen.[6]

How were responsibilities divided over your own internal team and external system integrator companies?

> We didn't work with a system integrator in the beginning. I had one very smart guy in my team, who's still one of our architects. Later, we had some help from a local integrator. Our success comes from a three-person team and the use of a product that was open and scalable. We built the initial template in-house. Nowadays, it should be easier to find a clever system integrator who has some predeveloped MES templates.
>
> So we started out with two people in the department in early 2004. As of January 2009, we have thirteen people internally employed, and we use a network of integrators in Denmark and Sweden. We contract them, and this extends our team to twenty-five people.
>
> Today, we do approximately 25 percent general development, and 75 percent directly delivering copy-paste solutions to the sites. That's where a company should get to, as quickly as possible, and it should be easier now than it was in our time. You should start small, with a scalable solution. The initial

6 Arne Svendsen (head of Manufacturing Services and Automation at Arla Foods Global IT), interview with the author, September 2008.

hardware and software costs (servers and licenses) should be less than 25 thousand dollars (programming and configuration excluded). Count on a month of engineering to get an area up and running. Of course, whether this is achievable depends on the current situation. For example, some PLCs are easy to integrate with the solution, but for other PLCs, you need an OPC driver and it becomes more complex.

How would you describe the difference between the end-user company's responsibilities and those of the system integrator?

We often have optimization projects to realize enhancements. Those are minor projects, with up to five hundred hours per project to get something working. But when we're going to build a new site, or extend a site, that's a different story. Sometimes we have a machine vendor who's responsible for the complete turnkey delivery of a production line. In that case, we'll tell him: "You bring in all the stainless steel and all the automation, and you implement the MES using our standard reports." That should be possible, but it's not an easy task. If we want a supplier to be responsible for this, then our template must be very clear. We'll have to bring enough knowledge to the supplier that they can build the solution like we would do it.

Currently, we hire people from system integrator companies into our team, and make sure they have the knowledge they need. We've built a big network, as big as possible. In Denmark and Sweden, there aren't that many people available yet who have a good knowledge of our MES product. We've been educating our integrators ourselves, along with our own people.

What kind of project phase terminology do you use?

We use a simple version of GAMP's V-model. We have a few standard documents. For example, we have a requirements specification that we're currently aligning with ISA-95. We don't develop a lot of design documentation. We have one global design guide document. This is the guide for all the programmers. Before you start a new project, the design guide specifies what's in the template and how the programmers should use it. Then we develop a small, fact-based requirements specification. We don't have any big test specifications at all.

This is probably also due to the fact that we're an internal supplier. We spend little time discussing something that's not working correctly. We don't do fixed-price projects, so instead of fighting about whether some kind of functionality is or isn't included, we choose to just develop the extra reports. This is quite different than the approach for our business applications. We have several hundred IBM people in India working on the business applications. In that case, the governance is very different. We spend much less time on governance and procedures in the Manufacturing IT team. In the IT projects, there's huge overhead, but in our projects we try to keep the governance down to a minimum. We have a project agreement document of about five pages. When we go to a site, we write just five pages of agreements about who has to do what, and when do we assess the project as being successful. So, for example, my team takes care of all the implementation hours, the hardware, the software, and so forth. But who's entering all the master data? That should be the customer on site, but they tend to forget that.

We don't spend a lot of time on testing. After the system goes live, our people will stay on site for about two months to resolve any issues. When we introduce new modules to the template, they're of course tested on test systems in our offices before taking them into production at the sites.

We don't write thick user requirements documents. We agree on what the user interfaces and reports should look like, and from which sources data will be collected. There's a template, so there isn't a lot of room for the users to ask for a solution different from the screens that are available in the standard template. They'll just have to accept that it looks different. So you need clear agreement with management at the site. They'll understand that it's cheaper and faster to use a standard solution.

Before we started it all, I had a lot of meetings with people at sites. This wasn't so much to understand what the requirements for the MES were; it was more of a sales effort. It's enough to study two or three sites per business group in order to understand the requirements for the template. Note that we more or less developed the template based on just the one cheese site. But it's been the vision from day one that we wanted to use the solution for as many sites as possible at Arla. An MES platform will work in the whole group, if it's good and flexible. There's no need to be afraid that there are too many differences; at the MES level, sites within one company are often very much alike, at least per business group.

What roles were there in the team?

The steering committee was the board of operations directors, one per business group, which was very high level in Arla. We spent half a year on template development, with small iterations. We got it into production as quickly as possible.

My advice is to work for at most a month before you get it out and give it to the users on site, so you can immediately evaluate its business value.

How do you take care of MES support at these sites?

Our team supports the sites and develops new functionality. In our team, there are two people working as internal consultants. They help sites with automation projects. They make sure that the MES template is followed, and that the global interface is used. The others work on further development of the MES template and implementation of the solution. We also spend quite some time on integration with the ERP system. We deliver a complete solution, including integration with Level 4.

For integration with SAP, we first used SAP XI, but a year ago we started using SAP MII. In Spain, the MII solution will go live in May 2009. We're using a very strict template, because we're implementing it in parallel at eight sites. So several teams are working in parallel, and they have to use the same template and project model.

We merged the user requirements specification, the functional specification, and the design specification from the GAMP V-model into a single document. It has a four-layer level structure. It also describes the integration between SAP and MES. Three teams are filling this in in parallel. By November 2008, we'll have the requirements specifications for eight sites. In the end, there'll be several kinds of documents available, such as requirements specifications, user manuals, and twenty-four-seven support documentation.

There'll be a team of seven people for support. They each take one week at a time. Our experience is that we have an average of three support calls per week for eight sites. In the spring of 2008, we rolled out a full MES implementation at a site. After going live, we had two weeks of troubleshooting, and then we went to the support phase. Two months after going live, we were down to one or two calls per month.

What about the MES rollout at different sites? Was it a "big bang" scenario? Why (or why not)?

We did the implementation at different sites in parallel. At some stages, we were working on up to eight or ten sites. It will always be necessary to do them in parallel, because you're not in control of when to implement at the sites. You're dependent on the sites' time and money. Every site has so many projects going on. You'll just have to fit in. You have to be able to switch over quickly.

Concerning the rollout of enhancements and new versions of the template, there is one point of concern. Not many of the MES tools have strong version control for software development, so it's a very difficult and manually driven process. It counts on human discipline to keep the implementation up to date. We may have a new version ready, but it can take many months to roll it out at the sites. And also: expect your MES supplier to come out with an upgrade every two years!

From day one, our team has delivered the engineering hours, because we want a global template for Arla Foods. The sites invest in the hardware, services, and software licenses that they need. Our team takes care of the discounts from suppliers, but the sites buy the functionality themselves. I pay for the annual upgrade out of my budget, because I want all the sites to use the same MES database version. We can't be three years behind, working on old versions. My team's decided to keep them on the same level. We also have a lot of work in testing a new version.

Do you pay attention to implementation support, in terms of training and coaching key users? How much time do you think should be spent on this?

Our biggest sites typically have one local technician, and he gets a three-day programming course on the MES platform plus a two-day course in SAP MII reporting, and we encourage them to make small enhancements. But small sites may not have their own technicians. So it's a mixed model. For the bigger sites, it's a matter of safety to have this kind of specialist on board. The smaller sites rely on our remote twenty-four-seven support, which can solve most problems at a distance. Of course, there's always a risk that these [local technicians] are going to build custom software that doesn't comply with the template.

We pay a lot of attention to personal networking and knowledge sharing. We organize workshops every three months, where [the participants] discuss their local projects and exchange knowledge. We spend time and money on this so people don't leave Arla Foods, and so they stay up to date. Bringing site people in for education in global standard platforms should always be free of charge and provide high-quality education.

As for training the MES end users, operator and supervisor training is a standard part of the project. Normally, two half-days of training will be sufficient. We don't have significant issues in MES user acceptance.

6.6 In Short...

Employing a multi-site MES template is not only cost efficient; it also makes unambiguous reporting possible, and thus benchmarking of the various sites. In many companies, multi-site MES templates are achievable, particularly by grouping plants with similar characteristics.

In preparing to create a template like this, it's a good idea to put together a working group containing representatives from the different plant groups, representatives from IT and Engineering, the future competence center leader, and one or more experts in the field of MES and vertical integration. This group will specify the multi-site MES template and choose the solution and the system integrator.

Vendors can only make a detailed price quote if—in addition to the MES template specification—they've also received sufficient information about the (pilot) site(s) where the implementation(s) will take place.

During the preparatory phase of your multi-site MES template project, the working group should already be thinking about how to set up the competence center that will manage the template. This competence center should comprise people with different skill sets and expertise. Preserving this knowledge and expertise is an important point of concern, especially if the MES template will be rolled out in multiple countries and the competence center employees will often be away from home.

While specifying, developing, implementing, and managing a multi-site template is a big job, it's still possible to approach it in a "lean" way, as in the Arla Foods example.

Now you know what MES is, what your company can gain from it, what to expect if you choose to implement an MES, and how to go about it if multiple sites are involved. You won't have failed to notice that in all those steps, the ISA-95 series plays an important role. What is ISA-95? When can you apply it? And what advantages does using this international standard convey? The next chapter answers these questions.

CHAPTER **7**

Why Should We Use ISA-95?

The ISA-95 series has been mentioned many times in this book. But just what is ISA-95? Why was this series developed? How can ISA-95 help your company more successfully prepare for and execute its MES projects? And how does it help the Manufacturing IT department save on interface costs?

7.1 Background on the Development of the ISA-95 Series

In the past twenty years, manufacturing companies have invested a lot of time and money in implementing centralized ERP systems on one hand, and automating their machines on the other. Then they wanted to integrate the two, but that turned out to be easier said than done. ERP system vendors told their customers, "Sure, you can create an interface to our system. Just provide the data according to our interface standard." But the traditional control system vendors also said, "Just provide the data according to *our* interface standard." The end user was left in the lurch. How could she ensure that the two systems from different vendors linked together seamlessly?

But it wasn't purely a technical challenge. Very many different people are involved in vertical integration projects. They each have a different view of manufacturing. The phrase *real time* means something very different to an IT person, for example, than it does to a process engineer. And yet these people have to work together to get the job done.

The question of what functionality still needed to be added in order to have the ERP system and the control systems connect properly, and whether that functionality belonged in the ERP or in a dedicated plant system, also led to constant discussions, which put such projects in danger of missing their deadlines. Pioneers discovered in the early nineties that it was essential to develop a clear, consistent view of the vertical integration issue.

Table 7.1 Overview of the parts of the ISA-95 series

The ISA-95 Series, Parts 1 through 5	Content Overview
ANSI/ISA-95.00.01-2000, Enterprise-Control System Integration, Part 1: Models and Terminology	This section contains models and terminology that provide an unambiguous view of MES and vertical integration. As such, it provides the foundation for good communication not only between people, but also between ERP and MES systems.
ANSI/ISA-95.00.02-2001, Enterprise-Control System Integration, Part 2: Object Model Attributes	This section contains models and tables of attributes. These specify in detail what information an ERP system should send to an MES system and vice versa, and what structure that information should have. Version 2 of Part 2 will also specify what information maintenance, laboratory, and warehouse management systems should exchange among themselves and with production and the ERP system.
ANSI/ISA-95.00.03-2005, Enterprise-Control System Integration, Part 3: Activity Models of Manufacturing Operations Management	Part 3 uses models and detailed textual explanation to clarify what activities take place within Level 3. These are, for example, the activities within the production department, the maintenance department, the laboratory, and the warehouse.
Part 4	At the time of this writing, the ISA95 committee had not yet begun Part 4. Part 4 will standardize the information flows within Level 3.
ANSI/ISA-95.00.05-2007, Enterprise-Control System Integration, Part 5: Business-to-Manufacturing Transactions	Part 5 specifies the Verbs (codified actions) systems should use to tell the receiving system what it should do with the data defined in Part 2. For example, "PROCESS Schedule" or "GET Equipment."

Against this backdrop, ISA began developing the ISA-95 series for Enterprise-Control System Integration at the end of the previous century. The ISA95 committee was founded. This committee comprised end users from diverse industries, ERP vendors, control system vendors, MES vendors, universities, system integrators, and consultants. They collected best practices and, based on these, put together a number of documents, each containing about a hundred pages. These documents contain agreements about what information an ERP system should send to an MES system, and vice versa.

The standard also clarifies how this information should be structured. You can compare it to a preprinted form, on which you just need to fill in the fields. For example, according to ISA-95, an ERP system should send a *Schedule* to the MES, and this schedule should contain (among other things) the *Start Time* and the *Priority*. And the MES should send back information about the *Production Performance*, containing (among other things) the *Material Consumed Actual* and the *Material Produced Actual*.

ISA-95 is a vendor-independent series of standards. The leading ERP and MES vendors support it, and now offer interfaces that are ISA-95 compliant. That's made integrating ERP and MES systems a less time-consuming and error-prone endeavor. Maintenance and reuse of these interfaces has also become simpler, allowing Manufacturing IT departments to save money.

7.2 The Enterprise Domain and the Control Domain

Before you can determine what information ERP and MES systems should exchange with one another, you first have to agree on what an ERP system is and what an MES system is. However, as we saw in chapter 2, ISA-95 doesn't use the abbreviations *ERP* and *MES*; rather, it employs the terms *Enterprise Domain* and *Control Domain*. In this way, it provides a vendor-independent view of the most logical place for certain functionality and information (see figure 7.1).[1]

1 The ISA95 committee used several basic principles to determine which functions belong in the Control Domain, such as "The function is critical to plant reliability" and "The information is needed by facility operators in order to perform their jobs."

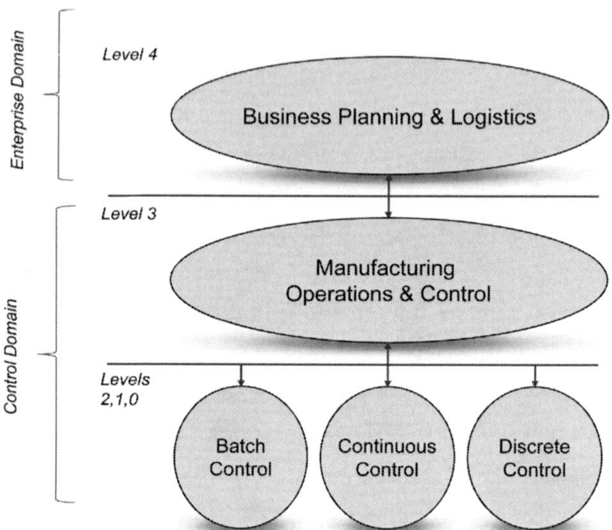

Figure 7.1 ISA-95 Functional hierarchy model (simplified)

The Enterprise Domain (Level 4) focuses on time spans of years, months, and weeks. Functions such as processing customer orders, purchasing materials, and sending invoices belong here.

The Control Domain contains several levels. Level 3 is the level of days, hours, and minutes. Activities such as recipe management, finite capacity scheduling, and production reporting take place here. Levels 2, 1, and 0 apply not only to batch processes, but also to continuous and discrete processes. These are monitored (Level 2: hours, minutes, seconds) and sensed and manipulated (Level 1); the actual production process takes place on Level 0.

ISA-95 doesn't prohibit ERP and MES vendors from working outside their traditional markets and offering functionality in the other domain. However, ISA-95 does make clear that there's a logical boundary between functions, and that the Enterprise Domain should be the owner of certain information, with the Control Domain the owner of other information. If a manufacturing company wants to reap the greatest possible benefit from the standard ISA-95 interfaces, it's wise to respect the ISA-95 boundary between Levels 3 and 4 when assigning functionality to the ERP system or the MES. If you do, it no longer matters very much whether a new plant to be purchased uses an MES system from vendor A or from vendor B. As long as that MES has an ISA-95-compliant interface, it can be relatively easily linked with the central ERP system.

Note that the standard contains more, and more detailed, models that help to more narrowly define the boundary between Levels 4 and 3, such as the models discussed in chapter 2 of this book.

7.3 B2MML (Business to Manufacturing Markup Language)

ISA-95 specifies the information that the Enterprise Domain and the Control Domain should exchange with one another, but it doesn't say what technology you should use to do so. After all, technology evolves rapidly. The ISA-95 agreements will still be valid even when new technologies replace the older ones.

At this time, XML is the most modern method for exchanging information. **XML** stands for *eXtensible Markup Language*. XML documents are text documents containing both data and metadata (see example 7.1).

Example 7.1 Part of an XML message

```
<MasterProductionPlan>
       <Number>0105200501095646</Number>
       <ReleaseDate>01-05-2005 T09.56.46.048</ReleaseDate>
       <Task>
              <TaskNumber>00010000431</TaskNumber>
              <Recipe>0010+000000000043000633</Recipe>
              <BeginTime>01-05-2005 T00.00.00</BeginTime>
              ...and so on...
       </Task>
</MasterProductionPlan>
```

The metadata are included in the form of tags. These are words enclosed in angle brackets, such as <BeginTime>. The data is between the tags. The surrounding tags assist us in understanding the meaning of the data. The big advantage of XML is that it defines itself. In other words: you can understand the information through the tags and the structure. The explanation of the information is contained in the same document as the information itself.

Example 7.2 An XML message converted to ISA-95 = B2MML

```
<ProductionSchedule>
        <ID>0105200501095646</ID>
        <PublishedDate>01-05-2005 T09.56.46.048</
        PublishedDate>
        <ProductionRequest>
                <ID>00010000431</ID>
<ProductProductionRuleID>0010+000000000043000633</
        ProductProductionRuleID>
                <StartTime>01-05-2005 T00.00.00</StartTime>
                ...and so on...
        </ProductionRequest>
</ProductionSchedule>
```

Instead of basing the data and metadata on application-specific language, you can also choose to use ISA-95, which contains the standard metadata and structure for production schedules, recipes, production capacity, and production results, among other things. In example 7.2, ISA-95 has been used as the basis for the structure and the metadata in an XML document. Note that the same data described in example 7.1 using the ERP application's metadata are described in example 7.2 using the standard metadata from ISA-95. Thus *MasterProductionPlan* has been translated to the ISA-95 term *ProductionSchedule*, and *BeginTime* has been translated to the ISA-95 term *StartTime*.

This combination of ISA-95 with XML is called **B2MML** (Business to Manufacturing Markup Language). B2MML is an XML implementation of the ISA-95 series. B2MML consists of several XML schemas that implement the data models from the ISA-95 series. They are free for download from the website of the Forum for Automation and Manufacturing Professionals (www.wbf.org). The schemas are the result of the valuable efforts of several members of the ISA95 committee, who set up a working group under the Forum for Automation and Manufacturing Professionals for developing and maintaining these ISA-95-specific XML schemas.

7.3.1 From the Trenches: Masterfoods Uses ISA-95 as Its Manufacturing IT Standard

The ISA-95 data models, upon which the Masterfoods scheduling system is based, have helped the vendor to develop a package suitable for scheduling processes in different types of plants. ISA-95 is, after all, an international standard, developed by parties that represent diverse industries. Gerard Alders, the person responsible for the information systems in Masterfoods' European plants, explains.[2]

> We don't want any things that have been specifically designed for Masterfoods. That could cause us problems in future releases. When we have special demands, we now ask ourselves, "Why do we have different requirements than the rest of the world?" So we've started to look much more critically at our own processes. In the past, we sometimes thought we had a monopoly on the truth, but that's no longer the case.

7.4 Possible Ways to Apply ISA-95

In practice, ISA-95 can be employed for diverse objectives. First, as already stated, it standardizes the interfaces between ERP and

2 Bianca Scholten, "Scheduling op basis van S95: Gedetailleerde planning van chocoladeproductie bij Masterfoods," *Automatie* (no.7, 2005), 10.

MES systems. In so doing, it makes vertical integration simpler. Manufacturing IT departments can develop a message once and then reuse it in different plants, between different kinds of systems. That ultimately makes a big difference in development and maintenance costs.

Moreover, ISA-95 is an important means of communication in MES project preparation, as has become clear in this book. The models in the standard help IT staff, engineers, plant managers, quality staff, maintenance staff, external consultants, and all others involved to better understand one another. For example, you can use them to clarify the scope of a project, to explain what functionality users can expect, to specify which interfaces will be automated, and so on.

ISA-95 offers not only models, but also terminology. It's *the* language for MES projects. MES and ERP vendors, system integrators, and consultants all speak this language. By writing user requirements specifications "in ISA-95," you ensure that your external vendor will understand you better and more quickly. It also has various advantages internally when different departments speak the same language. Multinationals such as Heineken, Arla Foods, and Nestlé have long since discovered this, and use ISA-95 as the standard language within their companies.

Finally, many Manufacturing IT departments also use the ISA-95 data models for developing software. The standard wasn't actually intended to be used this way; but here, too, it offers important advantages. The ISA-95 data models have been very well thought out. They're based on best practices. Why reinvent the wheel when you need to (for example) develop custom software?

7.4.1 From the Trenches: BAT Manufacturing's Use of ISA-95

Erwin Winkel, Senior Project Manager at BAT Manufacturing, describes the advantages that using ISA-95 has brought the company.[3]

> ISA-95 is an accepted standard within BAT Manufacturing. Since the initial phase of the project, we've used ISA-95. For example, we drafted the functional specifications based on the models and terminology in Part 1 of the standard.
>
> In the beginning, it took a lot of effort to convince people. They asked themselves, "What do we need that for? We know what we want, let's not make things so difficult." My argument was, however: we know what we want now, but not what we'll want later. If you base yourself on a standard, then it becomes easier to extend things, and you'll be more compatible in the future with packages based on the standard. ...
>
> The MES system's end users keep coming up with new desires, things they hadn't thought of before. Thanks to the ISA-95 structure, it's relatively easy to, say, add material qualities to the system. The effort required to make extensions is thus small. We also already see benefits in terms of maintenance. If employees go on vacation and others have to take over their work, it's relatively easy, because the standard is unambiguous.

7.5 In Summary

All in all, we can conclude that the ISA-95 series helps improve communication between people. It prevents misunderstandings and provides assistance in creating a clear definition of the company's

3 Bianca Scholten, "S95 toegepast bij eindgebruikers," Automatie (no. 10, 2003), 18.

processes and information flows, which is a prerequisite for automating them. ISA-95 also lays the foundation for the choice of which system can best house certain functionality, in order to reap the most possible benefit from standardized interfaces. That helps you limit the costs for vertical integration and interface maintenance.

For companies that want to build custom software themselves, to provide functionality missing from an MES, for example, ISA-95 offers well-thought-out data models. As a result, the quality of this software is less dependent on individual employees. Moreover, colleagues can more easily take over each other's tasks in cases of illness and vacation. In short: there's more than enough reason to adopt ISA-95 as *the* manufacturing IT standard for your company.

Now you have a first impression of what an MES has to offer your company, and how ISA-95 can help improve communication between people and systems in MES projects, thereby reducing the cost of vertical integration. I wish you good luck in your MES and vertical integration projects; may you quickly reap the rewards!

Bibliography

Bekkem, J. Van "Zegt het ene systeem tegen het andere; Akzo Nobel integreert SAP en Proficy." *Automatie*, number 6, 2007.

Dijkgraaf, W. and Spall, M. van *Start at the End with SMART Requirements.* Chaam: Synergio, 2007.

Huijs, L.W. *Manufacturing Execution Systems in Discrete Production Environments.* Unpublished manuscript, Eindhoven: Fontys University of Applied Sciences, 2002.

ISA. *ANSI/ISA–88.01–1995 Batch Control Part 1, Models and Terminology.* Research Triangle Park: ISA, 1995.

ISA. *ANSI/ISA–88.00.02–2001 Batch Control Part 2, Data Structures and Guidelines for Languages.* Research Triangle Park: ISA, 2001.

ISA. *ANSI/ISA–88.00.03–2003 Batch Control Part 3, General and Site Recipe Models and Representation.* Research Triangle Park: ISA, 2003.

ISA. *ANSI/ISA–88.00.04–2006 Batch Control Part 4, Batch Production Records.* Research Triangle Park: ISA, 2006.

ISA. *ANSI/ISA–95.00.01–2000 Enterprise-Control System Integration Part 1, Models and Terminology.* Research Triangle Park: ISA, 2000.

ISA. *ANSI/ISA–95.00.02–2001 Enterprise-Control System Integration Part 2, Object Model Attributes.* Research Triangle Park: ISA, 2001.

ISA. *ANSI/ISA–95.00.03–2005 Enterprise-Control System Integration Part 3, Activity Models of Manufacturing Operations Management.* Research Triangle Park: ISA, 2005.

ISA. *ANSI/ISA–95.00.05–2007 Enterprise-Control System Integration Part 5, Business to Manufacturing Transactions.* Research Triangle Park: ISA, 2007.

ISA. *ANSI/ISA–99.00.01–2007 Security for Industrial Automation and Control Systems, Part 1, Concepts, Terminology and Models.* Research Triangle Park: ISA, 2007.

MESA International. *MESA Metrics that Matter Guidebook and Framework.* MESA, 2006.

MESA International & Industry Directions Inc. *MESA Metrics that Matter; Uncovering KPIs that Justify Operational Improvements.* MESA, 2006.

MESA International *White Paper Number 2, MES Functionalities & MRP to MES Dataflow Possibilities.* MESA, 1997.

MESA International, *White Paper Number 4, MES Software Evaluation / Selection.* MESA, 1996.

MESA International *White Paper Number 6, MES Explained; A High Level Vision for Executives.* MESA, 1997.

Scholten, B. "IT or Engineering; Which of Them Should Support MES?" WBF, North American Conference, 2008.

Scholten, B. "Plant Dashboards; Veel fabrieken sturen nog via de achteruitkijkspiegel." *Automatie*, number 2, 2008.

Scholten, B. "Plant to Enterprise; MESA organiseert Europese conferentie in Utrecht." *Automatie*, number 9, 2007.

Scholten, B. "Productie als schakel in de keten; Business Process Management systemen dirigeren het orkest." *Automatie*, number 4, 2008.

Scholten, B. "S88 Stand van Zaken." *PT Industrieel Management*, number 2, 2005.

Scholten, B. "S95 integreert MES en ERP." *PT Industrieel Management*, number 10, 2005.

Scholten, B. "S95 toegepast door eindgebruikers." *Automatie*, number 10, 2003.

Scholten, B. "Scheduling op basis van ISA-95: Gedetailleerde planning van chocoladeproductie bij Masterfoods." *Automatie*, number 7, 2005.

Scholten, B. *The Road to Integration; A Guide to Applying the ISA-95 Standard in Manufacturing*. Research Triangle Park: ISA, 2007.

Scholten, B. "Van SAP tot PLC; Hoe bereik je 100% integratie?" *Automatie*, number 1, 2006.

Scholten, B. "Waarom investeren in MES en Plant Dashboards? MESA publiceert rapport 'Metrics that Matter'". *Automatie*, number 3, 2007.

Scholten, B. "Wat wil je, wat krijg je? Het schrijven van SMART requirements." *Automatie*, numbers 7 & 8, 2008.

Snoeij, J. *MES Product Survey*, Logica, 2008.

Svendsen, A. "*Bringing Business and Manufacturing IT Together.*" WBF European Conference, 2008.

Visser, R. "*MES Experience Survey.*" MESA Europe Conference, 2007.

Werckx, C. and Laer, C. van "*Approach to a Single MES Strategy Within a Global Environment.*" WBF European conference, 2008.

Glossary

21 CFR 11	21 Code of Federal Regulations Part 11: Electronic Records; Electronic Signatures.
APS	Advanced planning and scheduling.
B2MML	Business to Manufacturing Markup Language. XML schemas for exchanging information according to the object models defined in ISA-95. Chapter 7 of this book covers B2MML in more detail.
batch process	A process that leads to the production of finite quantities of material by subjecting quantities of input materials to an ordered set of processing activities over a finite period of time using one or more pieces of equipment.
BI	Business intelligence.
BOM	Bill of material. A listing of all the subassemblies, parts, and/or materials that are used in the production of a product. It includes the quantity of each material required to make a product.
BPMS	Business process management system.
continuous process	In a continuous process, materials are passed in a continuous flow through processing equipment. Once established in a steady operating state, the nature of the process is not dependent on the

length of time of operation. Start-ups, transitions and shutdowns do not usually contribute to achieving the desired process.

DCS Distributed control system.

discrete process In a discrete parts manufacturing process, products are classified into production lots that are based on common raw materials, production requirements and production histories. A specified quantity of products moves as a unit (group of parts) between work stations, and each part maintains its unique identity.

DS Design specification.

EAI Enterprise application integration.

EBITDA Earnings before interest, taxes, depreciation, and amortization.

ERP Enterprise resource planning.

FAT Factory acceptance test.

FDA United States Food and Drug Administration.

FS Functional specification.

FTE Full-time equivalent.

GAMP Good Automated Manufacturing Practice.

genealogy In manufacturing, genealogy refers to recording the changes made in every component of a product throughout its life.

GxP General term for good-practice quality guidelines and regulations used in many fields, including the pharmaceutical and food industries. The titles

	of these good-practice guidelines usually begin with *Good* and end in *Practice*, with the specific practice descriptor in between (e.g., GMP = Good Manufacturing Practice). GxP represents the abbreviations of these titles, where x (a common symbol for a variable) represents the specific descriptor. (Source: Wikipedia)
HMI	Human-machine interface.
hybrid process	A process that has characteristics of two or more process types: batch, discrete, and/or continuous.
ICT	Information and communication technology.
ISA	International Society of Automation (www.ISA.org).
ISA-88	ANSI/ISA-88 Batch Control standard.
ISA-95	ANSI/ISA-95 Enterprise-Control System Integration standard.
ISA-99	ANSI/ISA-99 Manufacturing and Control Systems Security standard.
ISPE	International Society for Pharmaceutical Engineering.
IT	Information technology.
IQ	Installation qualification.
JIT	Just-in-time.
KPI	Key performance indicator.
LIMS	Laboratory information management system.

master data Also called *reference data*, master data is any information that is considered to play a key role in the core operation of a business. Master data may include data about clients, personnel, material, products, equipment and more. Master data is typically shared by multiple users and groups across an organization and stored on different systems. (Source: Webopedia)

MES Manufacturing execution system; manufacturing enterprise solution.

MESA Manufacturing Enterprise Solutions Association International (www.MESA.org).

metadata Data about data. Metadata describe the meaning of data.

MOM Manufacturing operations management.

MPS Master production schedule.

OEE Overall equipment effectiveness.

OPC OLE for process control; open connectivity via open standards (a series of standards specifications).

OQ Operational qualification.

OSHA United States Occupational Safety and Health Administration.

Pareto analysis The use of a Pareto chart to identify which problems to focus upon, which makes a cause-and-effect analysis possible.

PC Personal computer.

PCS Process control system.

PLC	Programmable logic controller; a controller, usually with multiple inputs and outputs, that contains an alterable program.
PoC	Proof of concept.
PQ	Performance qualification.
Q&A	Questions and answers.
QA	Quality assurance.
QC	Quality control.
R&D	Research and development.
RFI	Request for information.
RFQ	Request for quotation.
ROA / RONA	Return on assets / return on net assets
ROI	Return on investment
root cause analysis	Class of problem-solving methods aimed at identifying the root causes of problems or events.
SAP	SAP AG, a business software applications and services provider.
SAP MII	SAP AG's manufacturing integration and intelligence solution.
SAP R/3	SAP AG's main enterprise resource planning software.
SAP XI	SAP AG's exchange infrastructure.
SAT	Site acceptance test.

SCADA Supervisory control and data acquisition; a generic name for a computerized system that is capable of gathering and processing data and applying operational controls over long distances.

SFC Sequential function chart.

SOA Service oriented architecture.

SOP Standard operating procedure.

SQL Structured Query Language, a standardized database language.

TA Technical automation.

URS User requirements specification.

WBF The Forum for Automation and Manufacturing Professionals, formerly known as the World Batch Forum (www.wbf.org).

WIP Work in process; work in progress.

WMS Warehouse management system.

XML Extensible Markup Language, an open standard designed to aid information systems in sharing structured data.

Appendix A:
MESA P2E Model Statement

The Manufacturing Enterprise Solution Association (or MESA) has been known as the source for models that explain and categorize the manufacturing plant information solutions space. The original MESA Model (see *MESA White Paper #06: MES Explained: A High Level Vision for Executives*, 1997) has been widely referenced in everything from textbooks to end-user specifications and industry analyst content. It was focused on Manufacturing Execution Systems (MES). This model was supplanted in 2004 with the new Collaborative MES (c-MES) model as described in *MESA White Paper #08: MESA's Next Generation Collaborative MES Model.* The c-MES model was focused on the concept of collaborative manufacturing, and how various systems worked together, including enterprise systems that did not traditionally fall into MESA's domain. This model and the associated definitions were also widely used by practitioners.

As the interest in manufacturing-related enterprise applications grew, so did MESA's membership base of high-level decision makers and strategists. MESA's focus began to shift into more of an enterprise view to provide guidance for how the manufacturing enterprise as a whole should implement initiatives involving disparate applications and business processes to achieve a strategic goal. This led MESA to its current focus on Strategic Initiatives, which are driven from the top down as efforts to improve the overall business. Ultimately, in any manufacturing organization, these strategic initiatives are what justify the cost required to implement and maintain software-based manufacturing solutions. In turn, these solutions provide the data and associated value targeted by the high-level strategy.

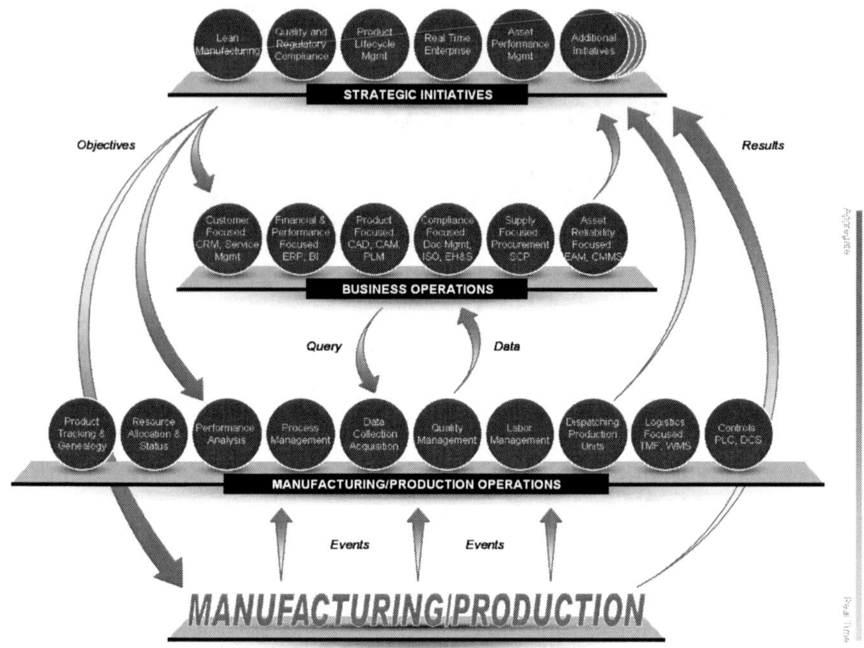

Today's MESA Model reflects this focus, and attempts to illustrate how enterprise strategy relates to operational practice. Today's model bears significant commonality to the models of the past. Both the Business Operations and Manufacturing/Production Operations tiers list common elements of past MESA models. One looks at this new model and can perceive that the Strategic Initiatives are highly dependent for sustainable success on the processes and technologies deployed in the Operations tier. The Strategic Initiatives are often the focus of corporate executives and board rooms in efforts to steer the enterprise in a general direction. To be successful in receiving the appropriate level of funding and support, a plant-to-enterprise (P2E) solution must continuously show how the investment helps achieve the objectives of the strategic initiatives and business/operations.

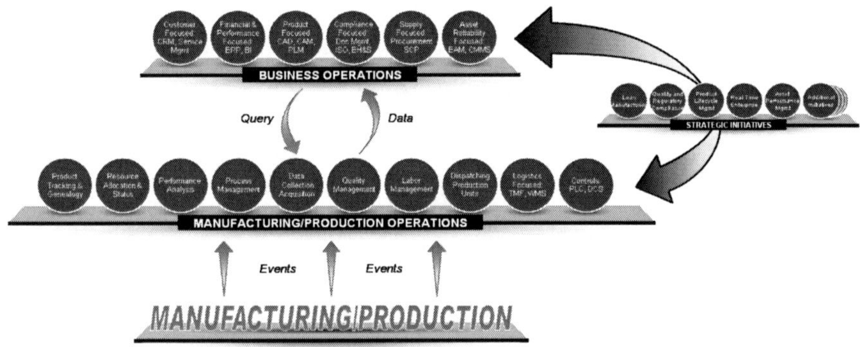

While this new model could have depicted only the three bottom tiers, it would have missed the primary forces that drive enterprise investment in these areas, achieving the objectives and results that are expected from true P2E connectivity. These results must show up as increased economic profit, improved customer satisfaction/relations, lean and agile operations, etc.

Consider the following:

The strategic initiatives are examples of common executive and board room endeavors to improve manufacturing enterprises. This is not an inclusive list; it's more like a "top five" list that will morph and change with external pressures. For example, sustainability became a significant initiative for many enterprises in 2008 due to rising energy costs. The expectation that there are many others is represented by the bubble furthest right that is titled, "Additional Initiatives." Executives are assigned initiatives and are subsequently required to show how they plan to achieve the objectives, and how success will be measured. In developing this model, MESA's objective was to help executives and managers identify the manufacturing IT strategies and best practices that support each of these initiatives. What systems are required? What are the risks and dependencies? How is success most often measured? What are the best practices associated with implementing systems required to support the initiative?

In 2008, MESA launched an aggressive program to leverage the knowledge of its broad based membership of end users, software vendors, and solution providers into a series of pragmatic guidebooks that target the most common strategic initiatives employed across a spectrum of industries. MESA's goal is to produce an illustrated compilation of findings that answer many of the most common questions posed when an initiative champion asks, "Why is that important?"

Every initiative can be decomposed into desired change and impact across a range of business and production systems and capabilities. Understanding the minimum required linkages and optimum potential relationships across these areas can mean the difference between true sustainable and measurable success versus the temporary illusion of it.

MESA strategic initiative guidebooks are dynamic in nature: they will be contributed to, updated, and revised to continually mirror the state of the industry. This will provide an ongoing evolution of P2E subject matter based on input from people on the front lines of the manufacturing information revolution.

Appendix B: **Table of Contents for the User Requirements Template for MES**

developed by the ISA-95 & MES Competence Center at TASK[24]

1. Management Summary
2. Introduction
   a. Purpose of the Document
   b. Project Planning Information
   c. Document Structure
3. Basic Description of the Company
   a. General Description
   b. Organizational Structure
   c. Physical Hierarchy
   d. Process Segments
   e. Current Application Architecture and Infrastructure
   f. Users
4. Business Drivers
5. Production Operations Management
   a. Product Definition Management
      i. Product Definitions
      ii. As Is
      iii. To Be
   b. Production Resource Management
      i. Production Resources
      ii. As Is
      iii. To Be

c. Detailed Production Scheduling
    i. Production Schedule
    ii. As Is
    iii. To Be

d. Production Dispatching
    iv. Production Dispatch List
    v. As Is
    vi. To Be

e. Production Execution Management
    vii. As Is
    viii. To Be

f. Production Data Collection
    ix. As Is
    x. To Be

g. Production Tracking
    xi. Reports
    xii. As Is
    xiii. To Be

h. Production Performance Analysis
    xiv. As Is
    xv. To Be

6. Maintenance

7. Quality Tests

8. Material and Energy Control and Product Inventory Control

9. Interfaces

    a. Interfaces with Level 4
    b. Interface with Maintenance Operations Management
    c. Interface with Quality Test Operations Management
    d. Interface with Inventory Operations Management
    e. Interfaces with Level 2

10. Other Activities

   a. Information Management
   b. Documentation Management
   c. Security Management
   d. Regulatory Compliance Management
   e. Configuration Management
   f. Incidents and Deviations Management

11. Miscellaneous
    a. Users and Workplaces
    b. System Administration Requirements
    c. System Documentation Requirements
    d. General System Requirements
    e. Language Requirements

12. Next Steps

13. Approval

14. Appendix A: Requirements List

15. Appendix B: Examples

16. Appendix C: ISA-95

17. Appendix D: Definitions and Abbreviations

18. Appendix E: Reference Documents

19. Appendix F: Legal Aspects

20. Appendix…

Index

Sponsors

Publication of this book is made possible in part by:

Care Automatisering develops world-class, user-friendly software that optimizes companies' complex, business-critical logistic processes and human, resource, and materials planning. Our award-winning detailed scheduling solution is fully based on the ISA-95 series so that your scheduled processes always reflect work-floor reality 100 percent.

For more information, visit: www.care-automatisering.nl

CORDYS

Cordys is a global provider of software for business process innovation and enterprise cloud orchestration. Manufacturers use Cordys to alter the way they innovate their Business Operations, to eliminate waste and support LEAN transformation initiatives. This is achieved by optimizing existing and new human workflows and system flows within one solution, while crossing the plant floor, front- and back-offices and value chains.

Cordys enables customers to achieve greater value from existing business infrastructure (ERP, PLM, CRM, SCM systems) by providing an advanced modeling, orchestration and process monitoring layer to optimize operational performance, to enhance the supply chain and create an agile enterprise. Cordys offers next-generation Business Process Management, Master Data Management, a Composite Application Framework, a scalable SOA Grid and Business Activity Monitoring in a single model-driven web-based platform. Cordys is headquartered in the Netherlands with offices around the globe.

For more information, visit www.cordys.com

De Clercq Solutions offers the IT platform 'Objective' to manufacturing companies aiming for operations execution excellence in order to increase their market competitiveness.

Objective was one of the first to combine MES- and WMS functionality into one solution, thereby covering the complete internal supply chain from an operational point of view. Based on the international ISA-95 series, Objective seamlessly integrates with all leading ERP and process automation systems.

Implementation templates are available for many industry-specific manufacturing and logistics processes, and related regulatory compliance.

For more information, visit www.declercqsolutions.eu

GE Fanuc Intelligent Platforms, a unit of GE Enterprise Solutions, delivers proven Operations Management Software solutions designed to reduce costs, increase efficiency and enhance profitability. GE Fanuc drives results on the plant floor and in corporate offices – empowering strategic business initiatives while delivering value to the plant, shift after shift, day after day.

The GE Fanuc Proficy™ Software suite includes SOA-enabled Workflow, Enterprise Integration, Enterprise Manufacturing Intelligence (EMI), MES, Quality & Compliance, HMI/SCADA, and more.

Whatever your biggest manufacturing challenges may be, GE Fanuc can help solve them – today, tomorrow and well into the future.

For more information, visit www.gefanuc.com

Respond fast to client orders, use resources most efficiently, maximize your OEE, and manage growing product variability and exceptions with MEScontrol.net.

MEScontrol.net is a modular Manufacturing Execution System that adheres to the principles of lean manufacturing in order to manage your production flow from A to Z.

MEScontrol.net is fully integrated with the plant-floor equipment, which is controlled in real-time by the centralized product definitions. The powerful scheduling engine guarantees the most efficient planning of orders based on resource availability. Continuous KPI measurements and OEE analyses provide valuable information to operators, managers and the system to maximize efficiency and uptime.

For more information, visit www.mescontrol.net

Rockwell Automation

One of the top challenges facing manufacturers is harnessing the power of their information — and that means faster, more accurate decisions that help drive out inefficiency and optimize manufacturing processes. Rockwell Automation is helping customers realize the benefits of convergence between information assets, the people who use them and the business processes they support. We help ensure that your plant operates as an integrated aspect of the enterprise and supply chain. Globalization is driving increased demands to improve quality, productivity, sustainability and regulatory compliance. We offer industry-leading solutions in discrete, hybrid and process industries around the globe.

For more information, visit www.RockwellSoftware.com

SAP is the world's leading provider of business software, offering applications and services that enable companies of all sizes across more than 25 industries to become best-run businesses. Today SAP is serving more than 82,000 customers in over 120 countries.

SAP Manufacturing is a comprehensive solution that synchronizes real-time production with demand volatility across global manufacturing assets, enables seamless integration between plant and enterprise systems and processes to drive responsive local execution. Industry leaders are achieving manufacturing excellence with SAP Manufacturing, SAP Manufacturing Integration and Intelligence (SAP MII), SAP Lean Planning and Operations, and SAP Manufacturing Execution.

For more information, visit www.SAP.com.

Wonderware is the market leader in real-time operations management industrial software which includes: Supervisory HMI, GeoSCADA, Production Management, MES, Performance Management, EMI, and integration with asset management, supply and demand chain and ERP applications. Wonderware has over 500,000 active software licenses in over 125,000 plants and facilities, and has customers in virtually every global industry — including Food & Beverage, Power, Water & Wastewater, Facilities, Transportation, Upstream Oil & Gas, Mining, Metals and other. Wonderware offers software solutions that tie together multiple fleets of plants and facilities, and enables new ways for customers, suppliers and producers to collaborate.

For more information, visit www.wonderware.com.